Fundamentals of Process Control Theory

An Independent Learning Module
from the
Instrument Society of America

Fundamentals of
PROCESS CONTROL THEORY

By Paul W. Murrill, Ph.D.

INSTRUMENT SOCIETY OF AMERICA

INSTRUMENT SOCIETY OF AMERICA
67 Alexander Drive
P.O. Box 12277
Research Triangle Park
North Carolina 27709

LIBRARY OF CONGRESS CATALOGING IN PUBLICATION DATA

LIBRARY OF CONGRESS
CATALOG CARD NO.: 80-84764

Murrill, Paul W.
  Fundamentals of Process Control Theory.

Research Triangle Park, N.C.: Instrument
Society of America

254 p.

8104   801107

ISBN: 0-87664-507-4

Editorial development and book design by Monarch International, Inc.

Fourth Printing 1988

Table of Contents

UNIT 12   **Automation System Concepts**

Preface

ISA's Independent Learning Modules

This is an Independent Learning Module (ILM) on the subject of *Fundamentals of Process Control Theory*; it is part of the ISA Series of Modules on Control Principles and Techniques.

The ILMs are the principal components of a major educational system designed primarily for independent self-study. This comprehensive learning system has been custom designed and created for the ISA to more fully educate people in the basic theories and technologies associated with applied instrumentation and control.

The ILM System is divided into several distinct sets of modules on closely related topics; such a set of individually related modules is called a series. The ILM System is composed of:

- The ISA Series of Modules on Control Principles and Techniques
- The ISA Series of Modules on Fundamental Instrumentation
- The ISA Series of Modules on Unit Process and Unit Operation Control

The principal components of a series are the individual ILMs such as this one. They are especially designed for independent self-study; no other texts or references are required. The unique format, style, and teaching techniques employed in the ILMs make them a powerful addition to any library.

ISA's goal is to publish at least four ILMs each year. There is obvious value in maintaining continuity within your personal set of ILMs; place a standing purchase order with ISA.

Comments about This Volume

This ILM on *Fundamentals of Process Control Theory* is designed to teach the basic principles of process automation and demonstrate the application of these principles in modern industrial practice. Some mathematics is necessary, of course, but efforts are made to prevent the mathematics from being a barrier to study by those without strong mathematical skills. The material is designed as an introductory or first-level course. A quick review of the Table of Contents will give an insight into the specific topics covered.

This textbook, like all of the ILM System, is designed for independent self-study. It is developed for the practicing engineer, first-line supervisor, or senior technician. In addition, college, university, and technical school students will find the material appropriate.

It is intended that this ILM be *both* theoretical and practical—that it show the basic concepts of process control theory and how these concepts are used in daily practice.

Unit 1:
Introduction and Overview

UNIT 1

Introduction and Overview

Welcome to ISA's Independent Learning Module *Fundamentals of Process Control Theory*. The first unit of this self-study program provides the information needed to take the course.

Learning Objectives — **When you have completed this unit, you should:**

    A. Understand the general organization of the course.

    B. Know the course objectives.

    C. Know how to proceed through the course.

1-1. Course Coverage

This is an introductory ILM on the fundamental principles of automatic process control. This course covers:

    A. The basic theoretical concepts of automatic process control.

    B. How these basic theoretical concepts are applied in modern industrial practice.

The scientific principles on which process control is based are unchanging. There is, however, considerable variation in the hardware available from vendors and in the application and practice of these control principles from one industry to another. The material presented in this course is generally oriented toward the modern practitioner in processing industries such as petroleum, petrochemical, chemical, pulp and paper, mining, power, drug and food processing.

This course will focus on process control and will not discuss individual measurement techniques (such as flow measurement or temperature measurement) except as measurement affects control.

No attempt is made in this ILM to provide exhaustive analysis of how to control a specific operation (such as heat exchangers or distillation columns), but the focus is more on the application of the general principles of control theory.

1-2. Purpose

The purpose of this ILM is to present in easily understood terms the basic theoretical principles of automatic process control and to illustrate and teach the usage of these principles in modern industrial applications. This is neither solely a theoretical course nor solely a practical course—it is both! The purpose is to show the theoretical concepts and principles in day-to-day commercial and industrial situations, and in doing so, to show that this theory is quite practical.

1-3. Audience and Prerequisites

This ILM is designed for those who want to work on their own and who want to gain a basic introductory understanding of automatic process control. The material will be useful to engineers, first-line supervisors, and senior technicians who are concerned with the application of process control. The course will also be helpful to students in technical schools, colleges, or universities who wish to gain some insight into the practical concepts of automatic process control.

No elaborate prerequisites are required to take this course, though an appreciation for industrial concerns and philosophies will be helpful. In addition, it is inevitable that some mathematics will be involved in particular parts of the presentation. It is not necessary for the student to be intimately familiar with such mathematics to appreciate the control concepts that are presented and applied. Quite often, mathematics becomes one of the barriers that prevents many persons from understanding and actually using process control theory; it is hoped that in this ILM such barriers will be minimized.

1-4. Study Material

This textbook is the only study material required in this course; it is one of ISA's new ILM System. It is an independent, stand-alone textbook that is uniquely and specifically designed for self-study.

There is contained in Appendix A a list of suggested readings to provide additional reference and study materials for the student. In addition, the student will find it most helpful to study the other ILMs that are available from ISA; these present a very broad range of specific applications of instrumentation and control material.

1-5. Organization and Sequence

This ILM is divided into twelve separate units. The next five units (Units 2-6) are designed to teach the student basic feedback control concepts and to teach the functional components that are used in modern industrial applications. Following these, there are two units (Units 7 and 8) to introduce the student to the dynamic behavior and tuning of process control systems. The next four units (Units 9-12) give the student an introduction to advanced control techniques and philosophies and to control system concepts. In addition, the significant impact of digital hardware is presented.

The method of instruction used in this ILM is self-study. Basically, you will work on your own in taking this course; you select the pace at which you learn best.

Each unit is designed in a consistent format with a set of specific *learning objectives* stated in the very beginning of the unit. Note these learning objectives carefully; the material which follows the learning objectives will teach to these objectives. The individual units contain example problems to illustrate specific concepts, and at the end of each unit you will find exercises to test your understanding of the material. All student exercises have solutions contained in Appendix D, against which you should check your solution.

This ILM belongs to you; it is yours to keep. We encourage you to make notes in the textbook and to take free advantage of the ample white space that is provided on every page for this specific purpose.

1-6. Course Objectives

When you have completed this entire ILM, you should:

A. Understand the basic theoretical concepts of feedback control.

B. Understand the functional role of the specific hardware components used in process control applications.

C. Have an appreciation of process dynamics and the tuning of industrial control systems.

D. Have an appreciation of advanced control techniques such as: cascade control, ratio control, dead time control, feedforward control, and multivariable control.

E. Understand the usage of digital processing capabilities in process control applications.

F. Have some appreciation of overall process control philosophies and strategies.

In addition to these overall course objectives, each unit in this ILM contains a specific set of learning objectives for that particular unit. These objectives are intended to help direct your study of that individual unit.

1-7. Course Length

The basic idea of the ISA System of ILMs is that students learn best if they proceed at their own personal pace. As a result, there will be significant variation in the amount of time taken by individual students to complete this ILM. Previous experience and personal capabilities will do much to vary the time, but most students will complete this course in 15 to 18 hours.

You are now ready to begin your detailed study of the basic concepts of automatic process control. Please proceed to Unit 2.

Unit 2:
Basic Control Concepts

UNIT 2

Basic Control Concepts

This unit introduces the basic concepts encountered in automatic process control; in addition, some of the basic terminology is presented.

Learning Objectives — **When you have completed this unit, you should:**

A. Be able to explain the meaning of the terms
   controlled quantities
   disturbances
   manipulated quantities.

B. Understand the basic concept of feedback control.

C. Understand the basic concept of feedforward control.

D. Have a general overview of process automation.

2-1. Control History

The first well-defined use of feedback control seems to have been Watt's application of the flyball governor to the steam engine in about 1775. As a matter of interest, most of the early applications and theoretical investigations were associated with governors and these usually were in industrial applications. Broader use of automatic control began to appear in the late 1920s and the first general theoretical treatment of automatic control was published in 1932. The growth in industrial usage has been steady and strong.

Many new technologies have been applied to process control hardware as the industrial use of automation techniques has developed and matured in the last 50 years. An important example was the application of digital computer capabilities to process control; in this case, process automation received a significant and very special boost in technology. Today, many industries allocate in excess of 10% of their plant investment capital outlays for instrumentation and control. This percentage has doubled over the past 30 years and shows no signs of diminishing.

The underlying theory of automatic control also has developed rapidly, and a firm and broad foundation of understanding has been created. Today's applications are based on this foundation. Many modern practitioners encounter difficulty, however, in the application of the well-defined mathematical theories of automatic process control. Much of this difficulty is quite natural, but much of the problem is due to the fact that often insufficient teaching is devoted to illustrating theoretical principles in day-to-day industrial applications. This ILM aims to alleviate much of this lack of understanding of the actual use of control theory in practice.

2-2. The Variables Involved

In understanding automatic process control, it is necessary first to fix in mind three important terms associated with any process. These are illustrated in Fig. 2-1. The *controlled quantities* or *controlled variables* are those streams or conditions which the practitioner wishes to control or to maintain at some desired level. These controlled quantities or controlled variables may be flow rates, levels, pressures, temperatures, compositions, etc. For each of these controlled variables, the practitioner also establishes some *desired value* or *setpoint* or *reference input*.

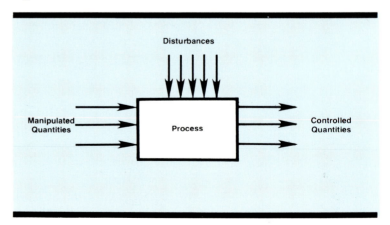

Fig. 2-1. The Variables Involved

For each controlled quantity, there is an associated *manipulated quantity* or *manipulated variable.* In process control this is usually a flowing stream, and in such cases the flow rate of the stream often is manipulated through the use of some control valve.

Disturbances enter the process and tend to drive the controlled quantities or controlled variables away from their desired or reference or setpoint conditions. The need then is for the automatic control system to adjust the manipulated quantities so that the setpoint value of the controlled quantity is maintained in spite of the effects of the disturbances. Also, the setpoint may be changed and then the manipulated variables will need to be changed to adjust the controlled quantity to its new desired value.

Fig. 2-2 shows a typical home heating system. In such a system, the controlled variable is room temperature. (Actually, if you wish to maintain creature comfort in the room, you typically control this through a variable that can be measured easily, such as temperature.) A number of forces cause room temperature to vary, for example, outside ambient temperature, the number of people in the room, the type of activity taking place in the room, etc. The automatic control system is designed to manipulate the fuel flow to the furnace in order to maintain room temperature at its desired value or setpoint in spite of the various disturbances.

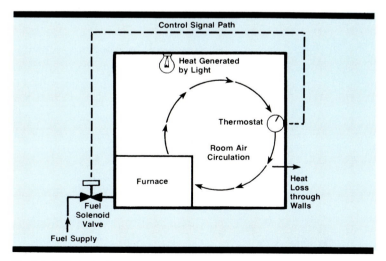

Fig. 2-2. A Home Heating System

2-3. Typical Manual Control

Before studying automatic process control, it is helpful to spend a moment or two reviewing a typical manual operation. This is illustrated in Fig. 2-3, a process with one controlled quantity. On the stream leaving the process, there is an indicator to provide operator information on the current actual value of the controlled variable. The operator is able to inspect this indicator visually and, as a result, to manipulate a flow into the process to achieve some desired value or setpoint of the controlled variable. The setpoint is, of course, in the operator's mind and the operator makes all of the control decisions. The problems inherent in such a simple manual operation are obvious.

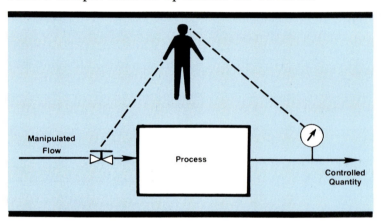

Fig. 2-3. Typical Manual Control

2-4. Feedback Control

The simplest way to automate the control of a process is through conventional feedback control. This widely used concept is illustrated in Fig. 2-4. Sensors or measuring devices are installed to measure the actual values of the controlled variables. These actual values are then transmitted to feedback control hardware, and this hardware makes an automatic comparison between the setpoints (or desired values) of the controlled variables and the measured (or actual) values of these same variables. Based on the difference (error) between the actual value and the desired values of the controlled variables, the feedback control hardware calculates signals that reflect the needed values of the manipulated variables. These are then transmitted automatically to adjusting devices (typically control valves) which manipulate inputs to the process.

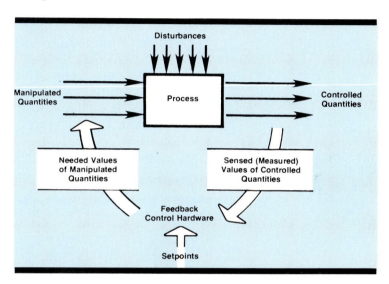

Fig. 2-4. Feedback Control Concept

The beauty of feedback control is that the designer does not need to know in advance exactly what disturbances will affect the process and, in addition, the designer does not need to know specific quantitative relationships between these disturbances and their ultimate effects on the controlled variables. The control hardware is used in a standard format, and all feedback control loops tend to reflect the general conceptual framework illustrated in Fig. 2-4. To a very significant extent, this standard pattern exists regardless of the specific nature of the process or the controlled variables involved. The particular hardware in a loop and the particular matching of one hardware piece to another is an important responsibility for the designer, but the overall control strategy is specifically defined. Such feedback control strategies are the simplest automatic process control techniques that can be developed, and this feedback control is typical of the vast majority of control technology used in industrial applications today.

2-5. Manual Feedforward Control

Feedforward control is much different in concept than feedback control. A manual implementation of feedforward control is illustrated in Fig. 2-5. In this situation, as a disturbance enters the process the operator observes an indication of the nature of the disturbance and, based on the entering disturbance, the operator adjusts the manipulated variable in such a manner as to prevent any ultimate change or variation in the controlled variable due to the disturbance. The conceptual improvement is apparent. Feedback control worked to eliminate errors but feedforward control operates to prevent errors from occurring in the first place. The appeal of feedforward control is obvious.

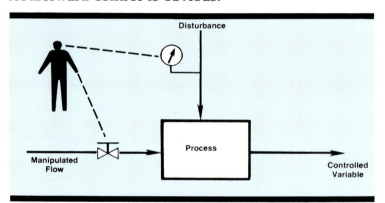

Fig. 2-5. Manual Feedforward Control

Feedforward control does escalate tremendously, however, the requirements of the practitioner. The practitioner must know in advance what disturbances will be entering the process, and there must be adequate provision to measure these disturbances. In addition, the operator must know specifically when and how to adjust the manipulated variable to compensate exactly for the effects of the disturbances. If the practitioner has this specific ability and if this ability is perfectly available, then the controlled variable will never vary from its desired value or setpoint. If the operator makes some mistake or does not anticipate all of the disturbances that might affect the process, then the controlled variable will deviate from its desired value and, in a pure feedforward control, an uncorrected error will exist.

2-6. Automatic Feedforward Control

Fig. 2-6 shows the general conceptual framework of automatic feedforward control. Disturbances are shown entering the process and there are available sensors to measure these disturbances. Based on these sensed or measured values of the disturbances, the feedforward controllers then calculate the needed values of the manipulated variables. Setpoints, of course, which represent the desired values of the controlled variables are provided to the feedforward controllers.

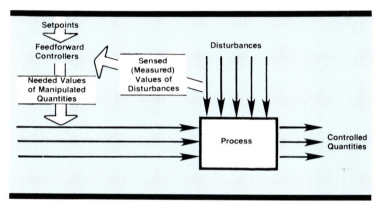

Fig. 2-6. Feedforward Control Concept

It is clear that the feedforward controllers must make very sophisticated calculations. These calculations must reflect an awareness and understanding of the exact effects that the disturbances will have on the controlled variables. With such understandings, the feedforward controllers are able then to calculate the exact amount of manipulated quantities required to compensate for the disturbances. These computations also imply specific understanding of the exact effects that the manipulated variables will have on the controlled variables. If all of these mathematical relationships are readily available, then the feedforward controllers can automatically compute the needed variation in manipulated flows to compensate for variation in disturbances. The escalation in required theoretical understanding is obvious. Feedforward control, while conceptually more appealing, significantly escalates the technical and engineering requirements of the designer and practitioner. As a result, feedforward control usually is reserved for only a very few of the most important loops within a plant. While the number of applications is small, their importance is quite significant.

Pure feedforward control rarely is encountered and the more common situation is for a process to have combined feedforward and feedback control loops. This will be illustrated in Unit 10 in which feedforward control is discussed in greater detail.

2-7.   Process Control *and* Process Management

Process automation is used to derive the maximum profitability from a process. In the material previously presented in this unit, there has been an implicit assumption that we knew the desired values (desired to achieve maximum profitability) for the controlled quantities. Once these desired values are known, techniques are automated to achieve and/or maintain these desired values or setpoints. Upon reflection, however, it can be seen that some of the most significant questions associated with the profitability of a process are the questions that must be answered to determine the desired values. This is basically the supervisory or management function and quite often it is left for the human operator to determine. But, in recent years, with the significant advances in process automation, many of these supervisory or management functions have themselves become automated, and the ability to achieve technological solutions and hardware answers of such management questions is a significant part of the modern control scene.

In a particular process, as the level of automation is increased, most of the initial steps involve using conventional process control (such as feedback control); but as the level of automation increases, more and more of the automation is associated with process management. This is illustrated in Fig. 2-7.

The combination of these two phenomena — process control and process management — must be reflected in our overall understanding and appreciation of process automation.

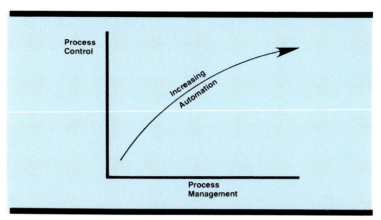

Fig. 2-7. Process Control *and* Management

Exercises:

2-1. *Consider an electric oven in a typical modern kitchen. Identify the controlled variable, the manipulated variable, and the disturbances.*

2-2. *Consider an automatic gas-fired, home hot water tank. Identify the controlled variable, the manipulated variable, and the disturbances.*

2-3. *Imagine you own a backyard swimming pool! Describe a manual control system to measure pH and to add an acidic solution to adjust pH. Define the controlled variable, the manipulated variable, and the disturbances.*

2-4. *Now automate the control of your swimming pool! Assume you have a tank of acid solution to pump into your pool to control pH; use feedback control.*

2-5. *"Cruise Control" of speed on an automobile is a good example of feedback control. Outline its operation in terms of feedback control.*

2-6. *Consider a gas-fired, home hot water tank being used in a house with heavy usage of hot water. This usage, of course, is the disturbance or load* on the tank. How could such a tank be controlled using feedforward control?

2-7.   For the hot water tank of Exercise 2-6, could you use combined feedback and feedforward control? How?

2-8.   Consider a "moon shot" by NASA as a preprogrammed feedforward control problem to send a rocket payload to the moon. Analyze such a shot in feedforward control terms. What is the significance of the mid-course correction?

2-9.   In many cases today, a home heating system's thermostat is coupled to a microprocessor so that the temperature setpoint may be managed in a manner to save energy. Analyze such a system in the terms of process control and process management.

Unit 3:
Functional Structure of Feedback Control

UNIT 3

Functional Structure of Feedback Control

The general concept of feedback control was presented in Unit 2. Now this general concept is reduced to a functional layout for a single feedback control loop.

Learning Objectives — **When you have completed this unit you should:**

  A. Understand the functional layout for a single feedback control loop.

  B. Be able to explain the components of block diagrams.

  C. Appreciate the mathematical structure of a single feedback control loop.

3-1. A Single Feedback Control Loop

Any given process will have a number of different controlled variables and for each controlled variable, there is an associated manipulated variable that must be chosen. Previous discussions have shown this only in the broadest and most general terms. Now a specific controlled variable is tied to a specific manipulated variable through the appropriate feedback control hardware. This is done in the manner illustrated in Fig. 3-1.

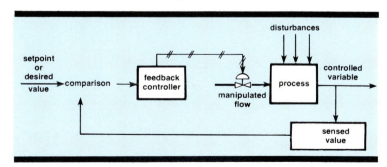

Fig. 3-1. A Single Feedback Loop

The controlled variable is sensed or measured through appropriate instrumentation and this sensed value of the controlled variable is then compared to the desired value of the controlled variable (the setpoint). The *difference* between these two (the error) is used as input to the feedback controller. This controller then calculates a signal to adjust the manipulated variable. Since the manipulated variable is normally a flow, the output of the feedback controller usually is a signal to a control valve, as illustrated in Fig. 3-1. While all of this is happening in a continuous fashion, disturbances may enter the process and tend to drive the controlled variable in one direction or another. The single manipulated variable is used to compensate for all such changes produced by the disturbances and, in addition, if there are changes in setpoint, the manipulated variable also is changed accordingly to produce the needed change in the controlled variable.

In a functional sense, all feedback control operates as illustrated in Fig. 3-1. Study this loop layout very carefully.

3-2. Block Diagrams

To have a consistent way of providing pictorial representation for control systems, it is useful to take advantage of block diagrams. Block diagrams are a simple, symbolic graphical tool commonly used in automatic control.

Block diagrams have two basic symbols; the first is a circle:

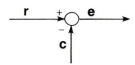

The arrows entering the circle and leaving the circle are not vectors. They do, however, represent variables and actually represent the flow of information. The head of each arrow has an algebraic sign associated with it, either plus or minus. If no sign is present, an implied plus is intended. The small circle is really a simple way to represent algebraic addition or subtraction. The symbol shown in the text above represents the algebraic equation $r - c = e$.

The other symbol of a block diagram is, in fact, a block with one arrow entering and one arrow leaving.

This is the way in which the algebraic operations of multiplication and division are symbolically presented. The output of the block is simply equal to whatever is contained within the block times the input. The block shown in the text above represents the equation $c = Ge$.

Block diagram symbols may be combined into networks. Shown in Fig. 3-2 is a block diagram of a very simple negative feedback control loop; it actually represents a combination of the two symbols shown earlier.

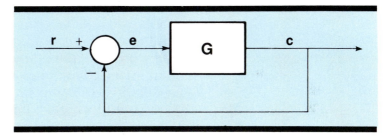

Fig. 3-2. Block Diagram of a Single Loop

Block diagrams are used consistently throughout this ILM to provide pictorial presentation of automatic control principles and applications.

3-3. The Functional Layout of a Feedback Loop

The general layout and structure of the single feedback control loop now needs to be expanded to more closely represent the way feedback control is used in practice. Such a functional layout is illustrated in Fig. 3-3. Basically, Fig. 3-3 is separated into two broad parts: The functional objectives which are accomplished *inside the controller case* and the balance of the process loop which is, of course, external to the controller case.

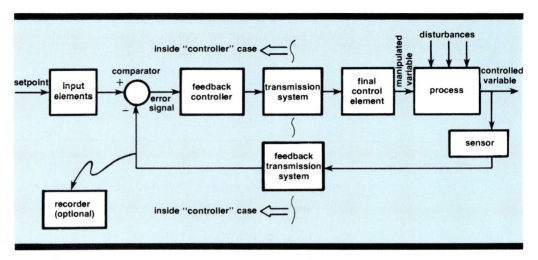

Fig. 3-3. The Functional Layout of a Feedback Loop

There must, of course, be provision for either the operator or some hardware to provide a setpoint to the control loop. This setpoint is the desired value of the controlled variable and will have the same dimensions as the controlled variable, e.g., if the controlled variable is gallons per minute, then the setpoint also will be gallons per minute. The input elements provide a functional conversion of the input setpoint signal into the operating mode of the controller e.g., millivolts, milliamps, psi air pressure, etc.

The controlled variable is actually measured by a sensor and the measured value of the controlled variable is transmitted back to the controller case. Inside the controller case is the *comparator.* This functionally important device actually compares—takes the algebraic difference between—the value of the setpoint (after conversion by the input elements) and the value of the variable transmitted back into the controller case to represent the controlled variable. The comparator, or *error detector,* is common to all feedback control systems. Note that this is negative feedback control, i.e., the signal fed back to the comparator is subtracted from the signal which indicates the setpoint. All applied feedback control is negative feedback control (positive feedback control is inherently unstable).

The error signal, which is the output of the

comparator, becomes the input to the feedback controller. Based on the error signal, the controller calculates a signal to the final control element—which is typically a control valve—and this in turn controls the manipulated variable input to the process.

Also shown with the feedback control loop of Fig. 3-3 is a recorder (for the controlled variable) which is optional.

In practice, many people refer to all of the items contained within the controller case as being the feedback controller. This broad general usage, while common, is not necessarily precise and will not be used throughout this ILM. Instead, it is more desirable to make functional distinctions among the various operations done around the feedback control loop and within the controller case.

3-4. Dynamic Components

The various blocks of a feedback control loop have different types of dynamic behavior, and it is most important that you gain insight into these process dynamics. This will be treated in more detail in subsequent units, but an introduction to the subject is now desirable.

Many of the individual components of the process control loop have no time-dependent behavior, i.e., there is no *lag* in their operation. When the input to the component changes, for all practical purposes the output changes instantaneously. This is illustrated below:

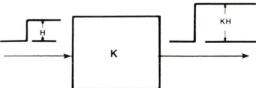

The type of component illustrated will be referred to as a nondynamic component. In effect, from a mathematical sense, it is algebraic in nature. The output changes instantaneously (for all practical purposes) when the input changes. In effect, the output is always proportional to the input and this proportionality constant will be referred to as the *sensitivity* or *gain* K of the component.

Many individual components illustrate dynamic characteristics. Typically, their output will lag behind any input. When such time-dependent behavior is encountered, it appears as illustrated below:

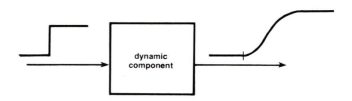

This type of dynamic response is typically encountered in the process itself, and to a lesser extent, in the control valve and in the sensor. The specific mathematical form of these dynamic lags will be discussed in more detail in later units.

3-5. Mathematical Model of a Loop

Earlier you looked at a functional layout of a feedback control loop. It is now appropriate that you understand this functional layout with some mathematical and dynamic insight into its behavior. The feedback control loop might be as illustrated in Fig. 3-4.

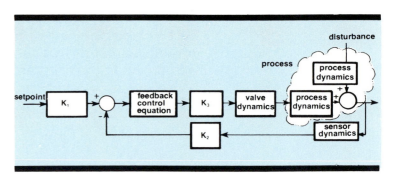

Fig. 3-4. The Mathematical Layout of a Feedback Loop

Note that the input elements are represented as a nondynamic component, i.e., when the setpoint changes, the signal to the comparator tends to change relatively instantaneously. The two transmission systems are also shown to be nondynamic. If they are well designed, they should not have significant time lags associated with them. (Sometimes in pneumatic transmission this becomes a problem and this will be discussed in a subsequent unit.)

To a lesser extent, both the sensor and the valve will have their own dynamics but, in the typical case, the dynamics (the lag) of these individual components will be less than that of the process itself.

All process loops are functionally the same and, in general, they follow the layouts that have been presented. Process dynamics vary significantly from one individual loop to another, and the practitioner must gain some appreciation and insight into the dynamics of an individual loop in order to design, install, and tune the loop to provide quality control. Quite often, students get distressed about the need to mathematically or quantitatively analyze dynamic performance. This frustration is understandable, but it does not eliminate the necessity of dealing with process dynamics. Process control is obviously needed only in situations that are changing, i.e., if nothing is changing, you do not need any control. Things that are changing are doing so with respect to time, and understanding their dynamic behavior is important. As a result, to understand process control, one must appreciate and understand process dynamic behavior.

As illustrated in Fig. 3-4, the process itself has dynamic lag all its own, and when disturbances enter the process, they will produce an effect on the controlled variable that is dynamic in nature. The same thing can be said of the manipulated flow as it enters the process.

Exercises:

> 3-1. *Fig. 2-2 shows a sketch of a simple home heating system. Develop a functional layout of the basic feedback control loop of this system.*

3-2. Given a simple level control system as shown below, develop a functional layout of the basic feedback control loop of this system.

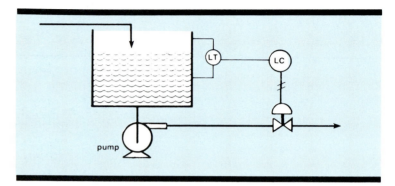

Fig. for Exercise 3-2

3-3. A temperature sensor has a range of 0-200°F and transmits a 3-15 psi air signal based on the measurement. What is its gain or sensitivity?

3-4. The input elements of an electronic controller accept a setpoint of 0-60 GPM and produce a 4-20 ma signal to the comparator. What is the gain or sensitivity of these input elements?

3-5. Estimate whether or not the dynamic characteristics of the following items of hardware would be significant in the total dynamics of a feedback control loop:

> the input elements in a controller
> an electronic transmission system
> a large pneumatically operated control valve
> a 2,000 ft pneumatic transmission system
> a pneumatic controller
> an electronic controller
> a bare thermocouple
> an orifice meter
> a chromatograph
> a thermocouple encased in a heavy thermowell.

3-6. In general, and for typical industrial applications, rank the following processes as to whether you would consider them as: a) responding very rapidly to input changes; b) responding at a moderate rate; or, c) responding very slowly.

liquid flow in a line
gaseous flow in a line
liquid level in a small tank
liquid pressure in an enclosed tank
gaseous pressure in a large tank
composition in a large distillation
    column
temperature in a liquid-filled tank
your co-workers at 5:00 p.m.

Unit 4:
Sensors and Transmission Systems

UNIT 4

Sensors and Transmission Systems

The quality of performance of a feedback control system is directly dependent on the quality of measurement of the controlled variable. In addition, this measured value must be transmitted to the controller in a timely fashion so that corrective action may be initiated. Also, the controlled output must be transmitted to the control valve. The purpose of this unit is to study the work of these measuring and transmission elements.

Learning Objectives — **When you have completed this unit, you should:**

A. Understand more fully the role and functioning of the sensors in a feedback control loop.

B. Be able to define *accuracy, precision,* and *sensitivity.*

C. Have insight as to what constitutes good dynamic behavior in a sensor.

D. Appreciate the characteristics of transmission systems.

4-1. The Sensor and the Transmitter

One of the most critical problems in designing and installing a feedback process control system is the specification of the sensing device that will obtain a continuous measurement of the controlled variable. In an operating sense, this measuring device not only provides a measurement of the controlled variable, but also produces *a change of variable.* The change of variable takes place since the controlled variable itself is not the actual signal that is transmitted back to the comparator. The transmitter produces an output signal whose steady-state value has a predetermined relationship to the controlled variable.

A transmitter is neither required from a measurement

nor a control point of view. Basically, the transmitter serves as an operating convenience by making the measurement data on the controlled variable available in a more centralized location, e.g., in a remote control room. Quite often, from a hardware viewpoint, the measurement function and the transmitter function are both incorporated into a single piece of hardware.

The principal controlled variables in process control systems are, in descending order of the frequency of their occurrence: *temperature, pressure, flow rate, composition,* and *liquid level.* Some of these variables, such as pressure, can be measured relatively directly while others, such as temperature, can be measured only indirectly.

One term which also should be defined is *transducer.* This is a general term for a device that receives information in the form of one or more physical quantities, modifies the information or its form or both, and sends an output signal. In a sense, depending upon the application involved, a transducer can be a primary measuring element (a sensor), a transmitter, a relay, a converter, or some other device.

You can appreciate that many of these terms—sensor, transmitter, converter, transducer—have broad and sometimes overlapping meanings and, quite often, the particular installation of a specific piece of hardware will dictate the appropriate descriptive term. Confusion can be minimized if the practitioner focuses principally on the functional usage involved.

4-2. Sensor Dynamics

It is important to gain an understanding of sensor dynamics. Quite often the speed of response of the primary measuring element is one of the most important factors affecting the operation of a feedback controller. Obviously, process control is continuous and dynamic, and the rate at which the controller can detect changes in the controlled variable will be critical to the overall operation of the system.

To gain some understanding of sensor dynamics, refer to Fig. 4-1 which shows a bare-bulb-type expansion thermometer. For analysis purposes, suppose that we immerse this bare bulb into an agitated constant temperature bath as shown.

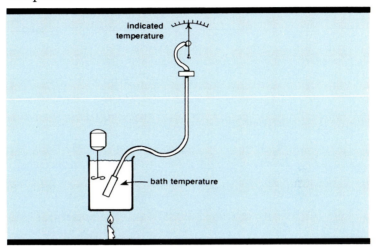

Fig. 4-1. Thermometer Experiment

The bare bulb will make the transition from ambient temperature to the temperature of the bath, and the thermometer needle might rise as is illustrated in Fig. 4-2. The curve shown in Fig. 4-2 is exponential and approaches the bath temperature gradually. A curve such as shown in Fig. 4-2 is referred to as a *response curve* and it gives experimental insight to the dynamics of this particular measuring device.

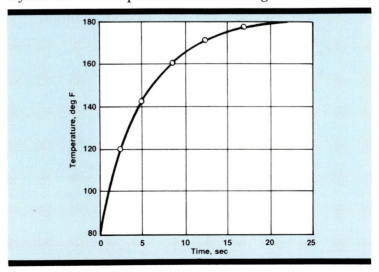

Fig. 4-2. Response of Bare Bulb Thermometer

Unit 7 will cover process dynamics in some detail. For the moment, the study of process dynamics is begun by defining a term used to characterize the dynamic behavior of a response curve such as shown in Fig. 4-2. This term is the *time constant* of the bulb and, basically, it is the time necessary for the response curve to reach 63.2% of its final value. In the illustration shown, the time constant for the bulb is approximately five seconds. The physical meaning of the term *time constant* will be explored in Unit 7, but for now we will use the term to make some quantitative comments about sensor dynamics.

It is the signal seen in Fig. 4-2 that is typically available for transmission, i.e., there will be some lag introduced into the feedback loop by the sensor. It is desirable that such lag be minimized wherever possible. Fast sensors make it possible for the controller to function in a timely manner. Sensors with large time constants are slow and degrade the overall operation of the feedback loop. The dynamic characteristics of sensors should be considered in their selection and installation.

4-3. Selection of Sensing Devices

Many questions must be considered before a specific means of measuring the controlled variable can be selected for a particular loop. There are no hard and fast rules for making such decisions, but there are a number of factors which must be considered:

A. What is the normal range over which the controlled variable might vary; are there extremes to this?

B. What accuracy, precision, and sensitivity are necessary? (These terms are defined in detail in the next section.)

C. What sensor dynamics are needed and available?

D. What reliability is required?

E. What are the costs involved—not simply the

purchase cost but also installation and operating costs?

F. Are there special installation problems, e.g., corrosive fluids, explosive mixtures, size and shape constraints, remote transmission questions, etc?

Obviously, with such a long list of important factors involved, the entire matter of selecting particular sensors for specific installations is very complex. As a result, as a part of the ILM System, separate ILMs are devoted to the subjects of how to measure temperature, how to measure flow rate, how to measure composition, etc. For study of these subjects, you are referred to these other ILMs published by ISA. For now, you will not explore in further detail the particular *how to* aspects of sensor selection; instead, these matters will be focused on only to the extent that they affect the basic performance and dynamics of the control loop.

4-4. Accuracy and Precision

Accuracy of a measurement is the term used to describe the closeness with which the measurement approaches the true value of the variable being measured. *Precision* is the reproducibility with which repeated measurements of the same variable can be made under identical conditions. In matters of process control, the latter characteristic is more important than accuracy, i.e., it is normally more desirable to measure a variable precisely than it is to have a high degree of absolute accuracy. The distinction between these two properties of measurement is shown in Fig. 4-3.

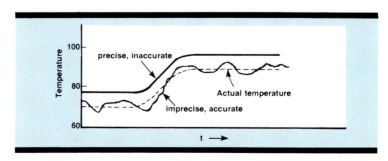

Fig. 4-3. Accuracy and Precision in a Temperature Measurement

The dashed curve is an indication of the actual temperature of a fluid. The upper measurement illustrates a precise but inaccurate instrument, while the lower measurement shows the measurement given by an imprecise but more accurate instrument. The first instrument has the greater error; the latter instrument shows the greater *drift*.

Practitioners make a distinction between two types of accuracy: *static* or *steady-state accuracy* and *dynamic accuracy*. Static accuracy is the closeness of approach to the true value of the variable when that true value is constant. Dynamic accuracy, on the other hand, is the closeness of approach of the measurement when the true value is changing. These terms are illustrated in Fig. 4-4. Clearly, the numerical value of the dynamic accuracy will depend on the nature of the dynamic change made by the true value of the variable being measured. In addition, the properties of the measuring system itself will have an effect. For process control systems, a practical specification of the time variation is ramp forcing, such as is shown in Fig. 4-4, and one practical designation of dynamic accuracy is the *dynamic error* resulting from ramp forcing.

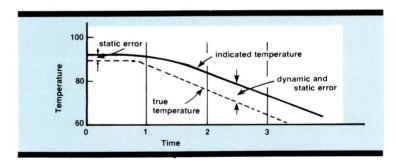

Fig. 4-4. Dynamic and Static Error

4-5. Sensitivity

Sensitivity of a measuring device is defined by the ratio of the output signal change to the change in measured variable. Clearly, the greater the output signal change for a given input change, the greater the sensitivity of the measuring element. Sensitivity is a steady-state ratio and, in effect, it is a steady-state *gain* of the element.

There is another kind of sensitivity which is very important in measuring systems. This sensitivity is defined as the smallest change in the measured variable which will produce a change in the output signal from the sensing element. In many physical systems, especially those which contain levers, linkages, and mechanical parts, there is a tendency for these moving parts to stick and to have some *free play*. The result is that small input signals may not produce any detectable output signal. Well-designed and well-constructed instruments are needed so that sensitivity will be high, and so that the control system has the ability to respond to small changes in the controlled variable.

4-6. Pneumatic Transmission

Often it is necessary to measure the value of the controlled variable at a location that is at a considerable distance from the controller. As a result, it is necessary to have a transmission system to get information to the controller and from the controller to the control valve. Typically this is done with either pneumatic or electronic systems. (Discussion of electronic digital transmission systems is reserved for Unit 11.) In this unit, transmission systems are discussed only from a control viewpoint; there is no detailed discussion of the precise hardware involved.

Pneumatic transmission may be used for distances up to several hundred feet. In the feedback transmission system, the controlled variable is measured and converted to an air pressure and a transmitter, in effect, sends this air pressure signal through a single tube to a receiver where it is transduced to a position or force for operation within the controller. A typical pneumatic transmission system is shown in Fig. 4-5. The controlled variable is sensed and converted to an air pressure and the measured pressure is often used as a pilot signal to an amplifier. (An amplifying pilot is often employed in pneumatic transmission systems in order to increase the air flow capacity of the transmitter.)

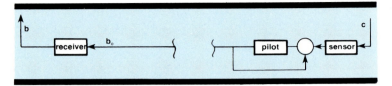

Fig. 4-5. A Pneumatic Transmission

The connecting tube carries the transmitted pressure to a receiver (located in the controller case.) This tube is almost always one-quarter inch in outside diameter and may be copper, aluminum, or plastic. The receiver is simply a pressure-gauge element, and the transmitted air pressure is converted into the movement of a bellows or diaphragm, i.e., pressure is transduced into a position or force that is used by the controller.

One of the most serious concerns in pneumatic transmission is the problem of transmitting a signal over significant distances because there is a lag associated with the transmission of the pressure signal through the long connecting tube. The tube has volume distributed along its length, and there is resistance to fluid flow through this tube. If a step change in input signal of 3 to 15 psi is made to a pneumatic transmission tube, then a typical output signal would appear as shown in Fig. 4-6.

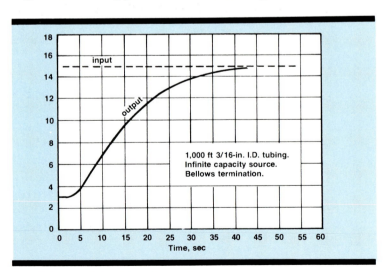

Fig. 4-6. Input and Output of a Pneumatic Transmission System

To model such dynamics for pneumatic transmission

systems, it is quite common to characterize them as being of the general empirical form shown in Fig. 4-7 in which the pneumatic transmission system is assumed to be adequately modeled as a *dead time* plus a *first-order lag.*

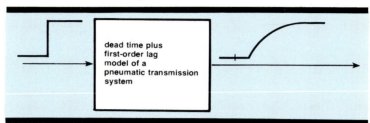

dead time plus
first-order lag
model of a
pneumatic transmission
system

Fig. 4-7. A Dead Time and First-Order Lag Model

Note that the first-order lag response is similar in nature to the dynamic response curve shown in Fig. 4-1 for the temperature measuring element. A dead time (θ seconds) is the time before any output signal is received. As before, the time constant (τ seconds) of the pneumatic transmission system is the time necessary (after the dead time) for this response curve to reach 63.2% of its final value. If a dead time plus a time constant is used as a model for the pneumatic transmission system, it is possible to produce data experimentally and then make a mathematical correlation of the responses. Fig. 4-8 shows such a correlation.

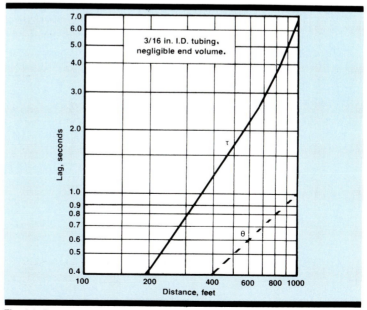

Fig. 4-8. Pneumatic Transmission Lag

Note that pneumatic transmissions may be employed for distances up to 350 feet if a time constant of about one second and a dead time of one-third of a second are acceptable. Fig. 4-8 also shows, for example, that transmission distances of up to about 1,000 feet are acceptable if a time constant of seven seconds and a dead time of one second are acceptable.

As distances increase, the speed of response of pneumatic transmission systems becomes a problem and alternate solutions are necessary.

4-7. Electronic Transmission

Electronic transmission systems have become much more practical and more common in the last two decades. Basically, such systems operate by transducing the controlled variable into an electric signal—usually a voltage or a current. Such systems have the advantage of virtually *instantaneous* response and therefore their dynamics do not become serious problems in process control applications. Because of their speed of response, such systems have the tremendous advantage of allowing nearly unlimited distances of transmission, either by wire or by radio linkage. The specific design and hardware layout of electronic transmission systems are not needed for a discussion of process control and, therefore, no elaborate presentation of the associated hardware is included in this ILM. Digital transmission systems will be discussed in Unit 11.

Exercises:

4-1. *The entire presentation of sensors and transmission systems was structured around the idea of feeding a signal back to the comparator. What would be involved (how would the hardware be different) if the process variable being measured was not a controlled quantity, but instead, was simply being sent to a central control room for indication and recording?*

4-2. *Refer to the immersion of the bare bulb in the bath of Fig. 4-1. How would the process response* curve be different if the thermometer bulb were not bare, but instead were encased in a protective thermowell?

4-3. A process chromatograph is used as a sensor to measure composition of a liquid stream. It operates on a discrete basis, i.e., it takes a sample and analyzes it, and establishes a signal for feedback purposes. It then takes a new sample and starts its cycle over again. What effect does this mode of operation have on loop operation as compared to the case of a continuous sensor such as, for example, a thermocouple or an orifice meter?

4-4. In Section 4-4, it was stated that sensor precision is more important than accuracy. How can this possibly be true?

4-5. How far can you transmit a pneumatic signal if you can tolerate a time constant lag of three seconds and a dead time of less than one second?

4-6. In a feedback control loop there are two transmission systems: One to feedback the signal indicating the controlled variable and one to send the controller output to the control valve. Are the process dynamic characteristics of one of these systems more critical than the other?

Unit 5:
Controllers

UNIT 5

Controllers

The feedback controller determines changes needed in the manipulated variable to compensate for disturbances that upset the process or for changes in setpoint. Understanding the controller's action is a necessity.

Learning Objectives — **When you complete this unit, you should:**

A. **Be able to explain proportional control action and its advantages and disadvantages.**

B. **Be able to explain reset (integral) action and its advantages and disadvantages.**

C. **Be able to explain rate (derivative) control action and its advantages and disadvantages.**

5-1. Controllers

The controller is a special-purpose calculator that uses the error signal from the comparator as its input or forcing function. It calculates the changes needed in the manipulated variable.

Quite often in hardware discussions, the controller case and everything inside that case is referred to as the controller. Within this case are many other functional elements of the feedback loop, e.g., the input elements, the comparator, the receiver from the feedback transmission system, and, quite often, a recorder.

Controllers usually are classified according to the chief source of power they use, i.e., electronic, pneumatic, mechanical, or hydraulic. All four of these types have response rates that are rapid enough for conventional process requirements, but in recent years the basic controls used in most process control applications have been either electronic or pneumatic. To a significant extent, they are competitive. Pneumatic actuators for control valves are much

cheaper and more satisfactory than electric actuators
and, therefore, instrument air is available in virtually
all industrial situations. Therefore, the intrinsic safety
and simplicity of pneumatic controls are important
advantages. However, the easy transmission and
manipulation of electronic signals gives *them*
significant advantages. This is especially true when
the instrumentation involves significant digital
computer hardware. As a result of these and many
other factors, the growth of electronic controllers has
been most significant during the last fifteen years.

Rather than concentrate extensively on the detailed
workings of specific pieces of hardware, the general
purpose of this unit will be to understand the various
modes of control action.

5-2. On-Off Control

All types of control action may be considered as either
continuous or discontinuous control. In one sense,
digital control is a special case of discontinuous
control and this will be treated separately in Unit 11.
When practitioners refer to discontinuous control
action, however, they usually are speaking of
two-position or multiposition control. Our primary
attention in this section will be on two-position
control.

Two-position control action, or *on-off control*, is
undoubtedly the most widely used type of control for
both industrial and domestic service. Many of us are
familiar with this type of control action since it is
used on most home heating systems and most
domestic hot water heaters.

Two-position control is a type of control action in
which the manipulated variable is quickly changed to
either a maximum or a minimum value, depending on
whether the controlled variable is greater or less than
the setpoint or some band width about the setpoint.
The minimum value of the manipulated variable is
usually zero (off).

The mechanism for generating on-off control is

usually a simple relay. In conventional practice, it is
not possible to build a device that is sensitive to the
sign of extremely small deviations—and more
importantly—it is not desirable to do so. Such an
excessively sensitive controller would undergo
needless wear and tear on its moving parts and
contacts and would keep the process very unsteady.
The solution in most commercial two-position
controllers is to establish a *dead zone* of about 0.5% to
2.0% of full range. The terms *differential gap* and
netural zone are often used synonymously with dead
zone. This dead zone, of course, straddles the
setpoint; no control action takes place when the
control variable lies within the dead zone itself. Fig.
5-1 shows what two-position control for a home
heating system might look like.

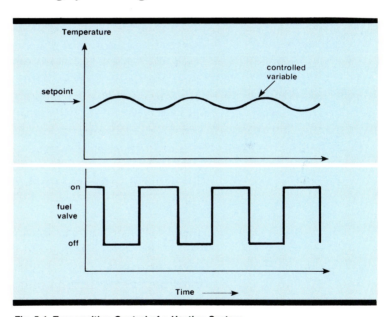

Fig. 5-1. Two-position Control of a Heating System

The instruments used for on-off control are cheap,
rugged, and virtually foolproof. On-off control is
inherently oscillatory in character, but for many
systems the amplitude of such oscillation of the
controlled variable can be quite small.

A variant of two-position control with differential gap
is three-position control, wherein the controller
responds with an intermediate output when the

controlled variable lies within the neutral zone. Fig. 5-2 illustrates this kind of control and its response characteristics.

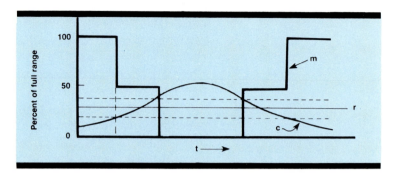

Fig. 5-2. Three-position Control

Commercial controllers with additional steps are available. Commercial controllers with as many as three intermediate output positions (this is effectively five-position control) are available, but the usage of such multiple-position control is not extensive.

5-3. Proportional Control Action

The basic continuous control mode is *proportional control* in which the controller output is algebraically proportional to the error input signal to the controller. The simple block diagram model of the controller shown below illustrates this:

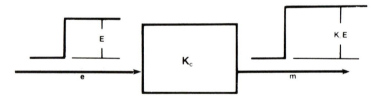

In this case, the controller output "m" is calculated as:

$$m = K_c e \qquad (5\text{-}1)$$

This equation is called the control *algorithm*.

Proportional control action is the simplest and most commonly encountered of all the continuous control modes. In effect, there is a continuous linear relationship between the controller input and output.

There are several additional names for proportional control, such as *correspondence control* (because of the linear correspondence of output to input), *droop control* (because of the droop or offset characteristic which will be discussed shortly), and *modulating control* (because of the proportional adjustment).

The *gain* of the controller is the term K_c; this is also referred to as the *proportional sensitivity* of the controller. It indicates the change in the manipulated variable per unit change in the error signal. In a very true sense, the proportional sensitivity or gain is an amplification and represents a parameter on a piece of actual hardware which must be adjusted by the operator, i.e., the gain is a knob to adjust.

The gain-adjusting mechanism on many industrial controllers is not expressed in terms of proportional sensitivity or gain but in terms of *proportional band* (PB). Proportional band is defined as the span of values of the input which corresponds to a full or complete change in the output. This is usually expressed as a percentage and is synonymous with *throttling band* or *throttling range*. It is related to proportional gain by:

(5-2)

$$PB = \frac{1}{K_c} \times 100$$

Since most controllers have a scale which indicates the value of the final controlled variable, the proportional band can be conveniently expressed as the range of values of the controlled variable which corresponds to the full operating range of the final control valve. This full operating range of the final control valve is often inferred to be the operation of a final control valve through a full stroke.

As a matter of practice, *wide bands* (high percentages of PB) correspond to less sensitive response and *narrow bands* (low percentages) correspond to more sensitive response. Several graphic means are used to illustrate the effects of varying proportional bands and examples of such are shown in Figs. 5-3 and 5-4.

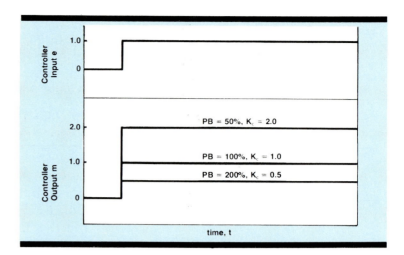

Fig. 5-3. Effect of Proportional Control on Controller Output

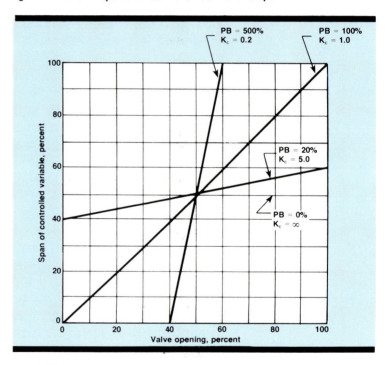

Fig. 5-4. Effect of Proportional Band on Valve Opening

Proportional control is quite simple and the easiest of the continuous controllers to tune, i.e., there is only one parameter to adjust. It also provides good stability, very rapid response, and dynamically is relatively stable.

Proportional control has one major disadvantage, however. At steady state, it exhibits *offset,* i.e., there is a difference at steady state between the desired value or setpoint and the actual value of the controlled variable. This is illustrated in Fig. 5-5.

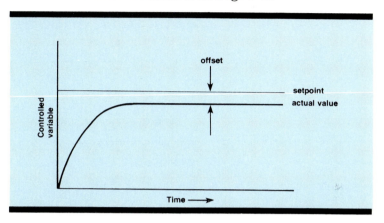

Fig. 5-5. Proportional Offset as Seen on a Feedback Recorder

5-4. Reset Control Action

Reset action is really an integration of the input error signal "e." In effect, this means that in reset action (often called *integral action*), the value of the manipulated variable "m" is changed at a rate proportional to the error "e." Thus, if the deviation is doubled over a previous value, the final control element is moved twice as fast. When the controlled variable is at the setpoint (0 deviation), the final control element remains stationary. In effect, this means that at steady state, when reset action is present, there can be no offset, i.e., the steady-state error must be zero.

Reset or integral control action usually is combined with proportional control action. The combination is termed *proportional-reset* or *proportional-integral action;* this is referred to as *PI control.* The combination is favorable in that some of the advantages of both types of control action are available. The basic control action for proportional-plus-integral action is as follows:

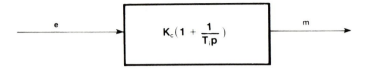

Where K_c is the *proportional gain*, T_i is the *integral time*, "e" is the error signal, "m" is controller output, and "p" implies the action of taking the derivative with respect to time, d/dt; therefore, 1/p implies integration with respect to time $\int \ldots dt$.*

While this type of block diagram representation for PI control has its advantages, it is desirable to expand it further, as shown in Fig. 5-6, which gives a detailed breakdown of the way a PI controller might function. In this particular case, the controller block diagram is separated into two parts, the upper part to show the proportional action—the lower part to illustrate the integral or reset action. A step input is provided to the controller and each of the control modes has its own characteristic output.

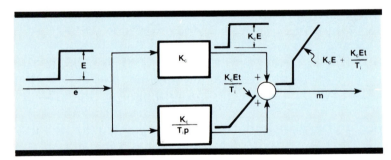

Fig. 5-6. Proportional-plus-Reset Controller's Response to a Step Change in Error

The total output of the controller is, of course, the sum of the output of the two individual modes. This also is illustrated in Fig. 5-6. It is seen that at the end of T_i units of time, the reset mode has tended to repeat (in magnitude of output) the proportional mode. On some controllers the adjustable parameter for the reset mode is T_i, the reset time, and on others it is the reciprocal of T_i, which is referred to as *repeats per minute.* Repeats per minute is also referred to as the *reset rate.*

*These terms, of course, are taken from standard calculus and for the reader who has not been introduced to such mathematics, some difficulties are inherent. With all apologies, the mathematics at this point cannot be avoided. p is the Heaviside operator d/dt.

The advantage of including the integral mode with the proportional mode is that the integral action eliminates offset. Typically there is some decreased stability due to the presence of the integral mode, i.e., the addition of the integral action makes the total loop slightly less stable. One significant exception to this is in liquid flow control. Liquid flow control loops are extremely fast and quite often tend to be very noisy. As a result, integral control is often added to the feedback controller in liquid flow control loops to provide a dampening or filtering action for the loop. Of course, the advantage of eliminating any offset is still present, but this is not the principal motivating factor in such cases.

Tuning a PI controller is more difficult than tuning a simpler proportional controller; now there are two separate tuning adjustments which must be made and each depends on the other. As a matter of fact, the difficulty of tuning a controller increases dramatically with the number of adjustments that must be made. The subject of tuning industrial controllers is covered extensively in Unit 8.

It is possible to use integral or reset action by itself without proportional control. This is not a common situation and will not be discussed further.

5-5. Rate Control Action

It is conceivable to have a control action that is based solely on the rate of change of the error signal "e." While this is theoretically possible, it is not practical because while the error might be huge, if it were unchanging, the controller output would be zero. Thus, *rate control* (often called *derivative control*) is usually found in combination with proportional control. The typical descriptive block diagram for a proportional-rate controller or PD controller is shown below:

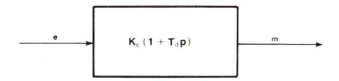

All of the terms are as defined earlier; in addition T_d is the derivative time and, as before, "p" implies the operation of taking the derivative with respect to time.

By adding derivative action to the controller, lead is added in the controller to compensate for lag around the loop. Almost any process has lag around the loop, and therefore the theoretical advantages of lead in the controller are appealing. It is quite a difficult control action to implement and adjust, however, and its usage is limited to cases in which there is an extensive amount of lag in the process. This often occurs with large temperature control problems.

In order to gain some insight into derivative action, refer to Fig. 5-7; the block diagram of the controller is broken into two parts to illustrate the separate action of the derivative mode and the proportional mode.

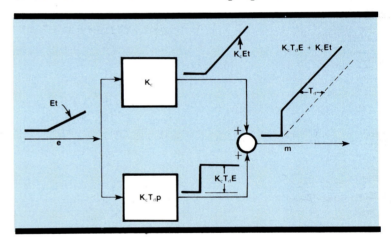

Fig. 5-7. The Output of a Proportional-plus-Derivative Controller for a Ramp Input

In this case, the input to the controller is shown to be a ramp change in error signal. By inspecting the combined output of both control modes, it is possible to see that the derivative time T_d, which is adjusted in the controller, is really an adjustment of the amount of lead that is introduced in the controller action.

The addition of rate control to the controller makes the loop more stable *if it is appropriately tuned*. Since the loop is more stable, the proportional gain may be higher and thus it can decrease offset above proportional action alone (but, of course, it does not eliminate offset).

5-6. PID Control

Proportional-plus-integral-plus-derivative control, or three-mode control, is the most sophisticated, continuous controller available in feedback loops. The typical block diagram for such a *PID controller* is shown below:

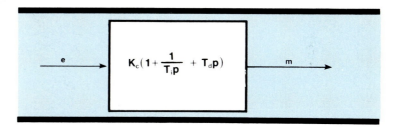

Fig. 5-6. Block Diagram for a PID Controller

In three-mode control, we have the most complex controller algorithm that is available routinely. It gives rapid response and exhibits no offset, but it is very difficult to tune—now there are three knobs to adjust. As a result, it is used only in a very small number of applications, and it often requires extensive and continuing adjustment to keep it properly tuned. It does offer very nice control when good tuning is implemented.

5-7. Summary

There is a need to summarize the different conventional control modes that have been presented, and Table 5-1 gives the basic descriptions of the various ones available. Table 5-2 illustrates the response action of these various control modes to different types of input forcing functions, i.e., to different changes in the input error signal. Table 5-3 gives a summary of the basic characteristics of these controllers.

Symbol	Description	Mathematic Expression
		One Mode
P	Proportional	$m = K_c e$
I	Integral (reset)	$m = \dfrac{1}{T_i} \int e\, dt$
		Two Mode
PI	Proportional-plus-integral	$m = K_c \left[e + \dfrac{1}{T_i} \int e\, dt \right]$
PD	Proportional-plus-derivative	$m = K_c \left[e + T_d \dfrac{d}{dt} e \right]$
		Three Mode
PID	Proportional-plus-integral-plus-derivative	$m = K_c \left[e + \dfrac{1}{T_i} \int e\, dt + T_d \dfrac{d}{dt} e \right]$

Table 5-1. Conventional Controller Modes

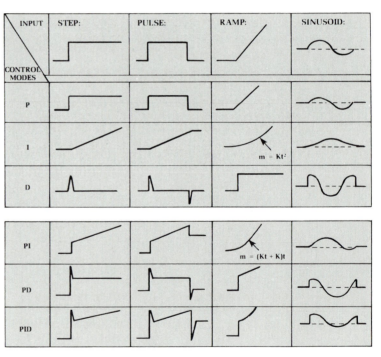

Table 5-2. Response of Controller Modes (shapes vary with actual values of K_c, T_i, T_d)

Two-Position:
  Inexpensive
  Extremely simple

Proportional:
  Simple
  Inherently stable when properly tuned
  Easy to tune
  Experiences offset at steady state

Proportional-plus-reset:
  No offset
  Better dynamic response than reset alone
  Possibilities exist for instability due to lag introduced

Proportional-plus-rate:
  Stable
  Less offset than proportional alone (use of higher K_c possible)
  Reduces lags, i.e., more rapid response

Proportional-plus-reset-plus-rate:
  Most complex
  Most expensive
  Rapid response
  No offset
  Difficult to tune
  Best control if properly tuned

Table 5-3. Characteristics of Controller Modes

Exercises:

5-1. For a proportional controller:
   (a) What gain corresponds to a proportional band of 200%?
   (b) What proportional band corresponds to a gain of 0.2?

5-2. Proportional-only controllers exhibit offset at steady state. How is this possible? (Offset implies an error; therefore, why does not the controller output eliminate the error?)

5-3. Throughout this unit, the error is shown as the input to the controller. Does it make any difference if the error is caused by a change in setpoint in one case or by a disturbance in another case?

5-4. Proportional control is often described by the equation $m = K_c e + M_r$ where M_r is an adjustable "manual reset" that can be used to bias the output to eliminate offset at a specific operating point. Will this eliminate offset at other operating points? Explain your answer.

5-5. It has been proposed that a controller might be designed to check the magnitude of the error and if the error is large, operate as a proportional-only controller; if the error is small, operate as a reset-only controller. What would be the advantage(s) of this type controller?

5-6. The "ideal" PID controller was described in Fig. 5-6:
$$m = K_c(1 + 1/T_i p + T_d p)e$$
but some have proposed that it would be better if it were described:
$$m = (K_p e + K_i/p)e + K_d p e$$
where K_P = proportional gain, equal to K_c
K_I = integral gain, equal to K_c/T_i
K_D = derivative gain, equal to $K_c T_d$
The latter form is referred to as noninteracting. What operating advantages, if any, would it possess?

5-7. All controllers include some filtering of the error signal before or within the rate mode. Why is this necessary?

5-8. Given the block diagram below of a PID controller with a step input in error as shown, draw the shape and give the equation of the output for each mode and the total controller output:

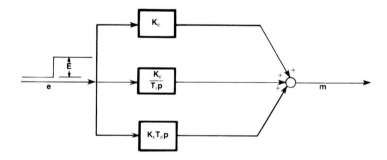

5-9. Repeat Exercise 5-8 for a ramp input of the form:

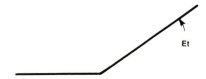

Unit 6:
Control Valves

UNIT 6

Control Valves

The output of the controller is a signal to the final control element which governs the control of the manipulated variable. In the vast majority of process control applications, the final control element is a valve and this most important piece of hardware merits understanding.

Learning Objectives — **When you complete this unit, you should:**

A. **Understand the purpose and use of control valves, actuators, and positioners.**

B. **Be able to define rangeability and turndown ratio.**

C. **Know the meaning and use of valve coefficients and appreciate sizing considerations.**

D. **Understand the factors influencing the dynamic behavior of control valves.**

6-1. Control Valves, Actuators and Positioners

For most process control systems, the final control element is a valve and these valves are typically driven by motors which are commonly called *actuators*. Actuators are classified on the basis of power source and valves are classified on the basis of the valve body style and flow characteristics. There is almost an unlimited number of hardware variations encountered in process control valves and actuators, and no attempt will be made in this unit to give an exhaustive presentation of the hardware itself. There is a specific ILM on control valves and the student is referred to this for detailed study. It would be helpful, however, to make a limited study of control valves. One is illustrated in Fig. 6-1; this is a sliding stem, single seat control valve body with a pneumatic actuator. It is shown here simply as an example.

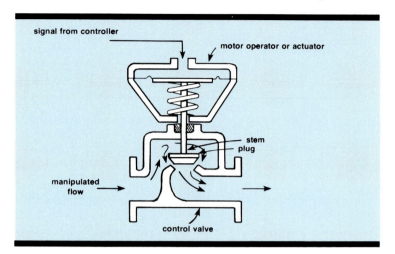

signal from controller

motor operator or actuator

stem
plug

manipulated
flow

control valve

Fig. 6-1. A Typical Sliding Stem Control Valve with a Pneumatic Actuator

It is convenient to classify control valve bodies as:
- A. linear stem motion control valves;
- B. rotary control valves;
- C. ball control valves.

The most popular of these categories will be illustrated in this unit.

The control valve actuator is used to translate the specific output signal from the controller into a position of a member exerting large power—this typically means that it drives the stem of the control valve. Control valve actuators may be pneumatic, electric, hydraulic, or manual. The most commonly used valve motors or actuators in the process industries are pneumatic and most of these have diaphragm motors. The diaphragm is spring loaded in opposition to the driving air pressure so that the valve stem position is proportional air pressure. Usually, the diaphragm is made of a rubber fabric or other limp material and is supported by a back-up plate. With diaphragm motors, the maximum available *stroke,* or stem travel, is usually two to three inches. For longer strokes, the valve actuator may be a double acting piston or it may be a rotary pneumatic device. Electric actuators are not too popular because of their cost and complexity; electro-hydraulic actuators are used primarily in areas were no operating air is available.

A *valve positioner* is actually a control valve accessory which transmits a loading pressure to an actuator to position a valve plug stem exactly as dictated by the signal from the controller. The use of a valve positioner often is desirable to improve both the dynamic and the static behavior of the valve. A valve positioner is typically an air relay which is used between the controller output and the valve diaphragm. It usually has a separate air supply and a feedback signal indicating stem position. The positioner acts to eliminate hysteresis, packing-box friction, valve plug unbalance due to pressure drop of the manipulated flow, and it assures the exact positioning of the valve stem in accordance with the controller output. The valve positioner also is useful in minimizing lag associated with the valve response.

6-2. Linear Stem Motion Control Valves

Linear stem motion control valves are like the valve shown in Fig. 6-1 in that the valve plug is positioned by the stem which slides through a packing gland. Linear stem motion control valves may be single seated, double seated, gate valves, etc. There are many advantages and disadvantages to the various types of valve plugs, and the ILM devoted strictly to control valves is a good source of understanding these various characteristics, advantages, and disadvantages.

The flow characteristics of linear stem motion control valves generally fall into three broad categories. These are illustrated in Fig. 6-2 and they are:

   A. *Decreasing sensitivity* type. In this case, the valve *sensitivity* (defined as the change in flow for a given change in valve position) decreases with increasing flow.

   B. *Linear* type. In this case the valve sensitivity is more or less constant throughout the flow range.

   C. *Increasing sensitivity* type. The most common

example is the *equal-percentage* type valve, and it derives its name from the fact that the valve sensitivity at any given flow rate is a constant percentage of that given flow rate. Other common terms used are logarithmic or parabolic type valve.

Of course, the individual flow characteristics of the wide range of hardware available will not always fall into such neat categories, but these do, at least, provide a systematic way for the practitioner to view the literally hundreds of valve trim arrangements that are available.

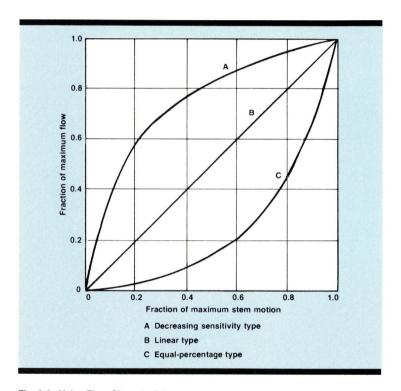

Fig. 6-2. Valve Flow Characteristcs

Linear stem motion control valves have many different body styles; the most common is in the form of a globe. These valves may be either single or double seated; examples are shown in Fig. 6-3. Single-seated valves are commonly employed for situations in which tight shut-off is required or in sizes of one inch or smaller.

Double-seated valves generally have leakage through the valve that is somewhat greater than in single-seated valves because it is virtually impossible to close the two ports simultaneously, especially when thermal expansion and other factors are considered. The advantage of double-seated bodies, however, is that the hydrostatic effect of the fluid pressure acting on each of the two seats will tend to cancel out and much less actuator force is necessary.

Linear stem motion valves are also used in three-way valve bodies where the control valve may be used to divert a stream or to combine streams. Linear stem motion valves are also commonly encountered in angle valve situations; in such cases the valve body is typically single seated. In addition, linear stem travel valves are built in Y-style bodies, in split-body styles, and cage styles. Cage valves usually are designed for easy removal of the valve trim to facilitate maintenance and replacement, if necessary. Linear stem motion valves are also used in sliding gate valve application and in a number of similar, special applications.

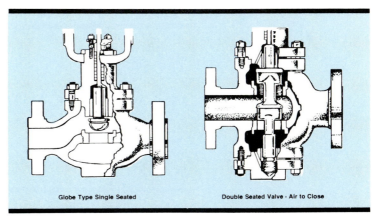

Globe Type Single Seated   Double Seated Valve - Air to Close

6-3. Linear Stem Motion Control Valves (Courtesy Masoneilan International, Inc.)

6-3. Rotary Control Valves

Rotary shaft control valves have enjoyed a substantial increase in usage in recent years. Their advantages are low weight, simplicity of design, relatively high flow rates, more reliable and friction-free packing, and relatively low initial cost. They cannot usually be used in sizes below 1".

The most common rotary shaft control valve is the butterfly valve, which is illustrated in Fig. 6-4. Butterfly valves are used in sizes from 2" through 36", or larger. They are often used in applications involving large flows at high static pressures but with limited pressure drop availability. Properly selected, the butterfly valve offers the advantages of low cost, light-weight simplicity, and saving space. It also exhibits good flow control characteristics.

One type of rotating shaft control valve of particular usefulness is the eccentric cylindrical plug valve, which is actually a modification of the plug cock widely used for shut-off service. Its relative capacity is high and its cost is low. It is especially useful on services involving corrosive fluids, viscous liquids, or suspended solids.

Also illustrated in Fig. 6-4 is the eccentric-rotating plug valve. This type can serve in a large percentage of all process control requirements.

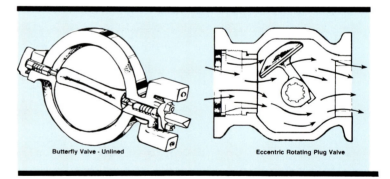

Fig. 6-4. Rotary Shaft Control Valves (Courtesy Masoneilan International, Inc.)

6-4. Ball Valves

The usage of ball valves for control purposes has

grown rapidly since the early 1960s, although ball valves as such are much older. The ball valve has historically been used principally as a tight shut-off hand valve. In recent years, however, ball valves have been automated for control purposes and have shown excellent rangeability and excellent suitability to handle slurries.

There are two distinctly different types of ball control valves. One involves a ball (a complete sphere) with a waterway through it; this is generally referred to as the full ball type. The second type is developed along the lines of the concentric plug valve and utilizes a hollowed out spherical segment, or partial ball, which is supported by shafts. These are referred to as characterized ball valves and are illustrated in Fig. 6-5.

Both types of ball valves are quarter-turned rotary valves. The characterized ball valve is basically a simple segment of a sphere which forms a crescent-shaped flow path which produces flows of equal-percentage character for high capacity designs, and a linear characteristic for low capacity designs. To obtain other characteristics, the leading edge of the segment can be contoured, as in the V-notch design and parabolic port designs.

In the full ball valve design, the shape of the control orifice goes from circular to elliptical as the ball is moved from its open to its closed position. This gives an essentially equal-percentage behavior. Changing the shape of the waterway in a full ball valve is not a practical approach to changing the valve's characteristic, and when such a goal is desired, it is often accomplished by using interchangeable cams in the positioner.

Ball valves have the highest flow capacity of any commonly used control valves. They are useful whenever slurries are involved; they also provide tight shut-off.

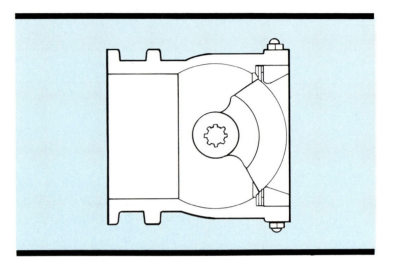

Fig. 6-5. Cross-Section of a Characterized Ball Control Valve (Courtesy Masoneilan, Inc.)

6-5. Control Valve Characteristics

Control valves are characterized by their rangeability, their turn-down, and their flow behavior.

Rangeability is the ratio of the maximum controllable flow through the valve to the minimum controllable flow through the valve. In typical operations, control valves do not close off entirely because to do so might damage the valve seat or cause the valve to stick. Flow in the closed position, therefore, is typically 2% - 4% of the maximum flow. This corresponds to a rangeability of fifty to twenty-five. Linear stem motion control valves show a rangeability of twenty to seventy.

Turndown is defined as the ratio of the normal maximum flow through the valve to the minimum controllable flow. A practical rule of thumb is that a control valve should be sized so that the maximum flow under operating conditions is approximately

70% of the maximum possible flow. Thus, the turndown ordinarily is about 70% of the rangeability.

Flow through a control valve depends not only on the extent to which the valve is open, i.e., on the *travel*, but also on the pressure drop across the valve. The next section of this unit gives specific insight into the broad question of the relationship between flow and pressure drop for a given valve opening.

6-6. Control Valve Selection and Sizing

There are many important factors to be considered in control valve selection; these include:

A. The rangeability of the process and the maximum specific range of flows required by the process.

B. The range of operating loads to be controlled.

C. The available pressure drops at the valve location at maximum flow and at minimum flow.

D. The nature and condition of the fluid flowing through the valve.

Control valve rangeability must exceed by some reasonable safety margin the rangeability of the process.

Not only must control valves have sufficient rangeability, but also they must be able to moderate flows which are well below the minimum requirement of the process. In addition, the maximum flow rate through the valve must exceed the peak requirement of the process.

For many process plants, the normal range of operating loads is narrow and, in such cases, the control valves maintain very nearly constant flow rates.

Valves selected to operate at 60% - 70% of full capacity are normally used in such applications. There are many process units, however, in which the operating conditions vary considerably, i.e., the load varies significantly. In such conditions, equal-percentage flow characteristics are often used, since they provide a more constant fractional sensitivity at all operating levels.

The nature and condition of the fluid have obvious effects on valve selection.

The entire question of valve pressure drop enters into not only the sizing of the valve but also the nature of the dynamic performance of the valve. The proper size of a control valve is most important to the overall operation of the feedback control system. If the valve is oversized, it tends to operate only slightly open and the minimum controllable flow is too large. In addition, the lower part of the flow characteristic curve is quite often nonuniform in shape. On the other hand, if the valve is undersized, the maximum flow needed may not be obtainable.

In sizing control valves, it is standard practice to combine many terms from the basic orifice equation into the following general relationship for incompressible liquids:

$$m_L = C_v \sqrt{\frac{\Delta P}{G}}$$

where:

m_L = liquid flow rate at the conditions for which specific gravity G is taken, gals per min (GPM)

G = specific gravity of liquid (referred to water) at either flowing or standard conditions

C_v = valve coefficient

ΔP = pressure differential, psi

The valve coefficient C_v is defined as the flow rate of water in gallons per minute provided by a pressure differential of 1.0 psi through a fully opened control valve. The size coefficient for any control valve must be determined by actual test. The valve coefficient for a control valve of the linear stem travel type is very approximately equal to the square of the nominal valve size multiplied by ten.

For the flow of compressible fluids, the valve coefficient is also employed with suitable conversion factors. The form of the equation is an approximation to the complete isentropic flow equation:

$$m_g = 63.3 \, C_v Y \sqrt{\Delta P \gamma_1}$$

where:

m_g = flow rate of gas, lbs per hour

C_v = valve coefficient (obtained for liquids)

ΔP = pressure differential across valve, psi

γ_1 = specific weight, lb /ft³, at upstream conditions

Y = expansion factor, ratio of flow coefficient for a gas to that for a liquid at the same Reynolds number, varies from 0.667 to 1.0

The entire question of determining control valve size, flow rate, or pressure differential is made in most industrial applications by the use of special-purpose calculators or special-purpose computer programs. Sizing also is done through the use of nomographs. In actual practice, the sizing of control valves is more complex than is illustrated here; the equations shown here do illustrate the basic principles involved, however. For specific details on the sizing of valves, the student is referred to ISA's separate ILM on control valve selection and sizing.

6-7. Control Valve Dynamic Performance

It is most important that you gain a specific understanding of the dynamic characteristics of a control valve installed as a part of a total piping network. The concepts involved significantly affect the valves' operating characteristics and the total performance of the feedback control system.

In Fig. 6-6 is shown an example piping network. In this case, a centrifugal pump takes a suction on a feed tank and pumps a liquid stream through a preheater and into a distillation column. There is a liquid flow control loop serving to control the actual liquid flow rate into the distillation column.

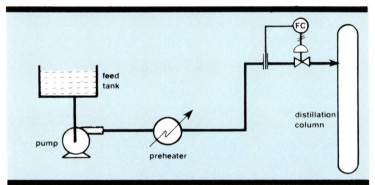

Fig. 6-6 Example Piping Network

If you closely inspect the total distribution of head loss throughout this piping network, you will get some insight into the dynamic characteristics of the valve. The centrifugal pump might have a total characteristic head curve as is shown in Fig. 6-7.

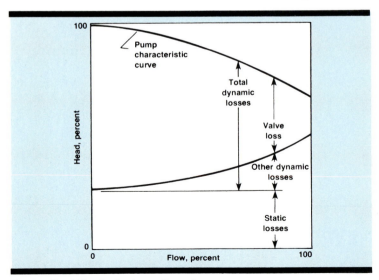

Fig. 6-7. Head Distribution in the Network

This pump output head is "consumed" by a number of different types of pressure drops. Many of these are *static* in nature and independent of the amount of fluid flowing through the piping network. One example of such static loss is the potential energy required to raise the liquid to the feedpoint on the column; a second static loss is illustrated by the operating pressure of the column itself which must be overcome by the inlet liquid feed.

In addition to these static losses, there are many *dynamic losses* involved in the piping network, other than the losses associated with the control valve. These additional dynamic losses tend to increase as the square of the flow rate through the piping network. These losses are the friction losses in the piping, flow through the preheater, flow through the orifice plate, etc. After the static and dynamic losses, all of the other head losses of the flowing fluid must be consumed across the control valve.

In a very real sense, the control valve is a bottleneck. It is a bottleneck that is designed and installed on purpose. If it is not the bottleneck, it is not the control point. As flow increases and a smaller percentage of the total dynamic drop occurs across the valve, then the piping network and not the valve tends to become the bottleneck.

When most valves are sold, the practitioner obtains a single characteristic curve for the valve. This is typically a curve such as shown in Fig. 6-2, which illustrated the three basic types of sliding stem control valves. These basic curves are representative of the situation in which the total dynamic drop occurs across the control valve. In actuality, for every single control valve, there is a whole family of control valve sensitivity curves, and the plotting parameter in the family of such curves is the percentage of the total dynamic drop which curves across the control valve. This is illustrated in Fig. 6-8 which shows a family of curves for a typical equal-percentage valve and for a typical linear valve. Note that if you have less than the total dynamic drop occurring across the control valve, the characteristic sensitivity of the valve may be significantly altered. These questions have remarkable impact on the tuning of process control systems and this will be discussed in Unit 8.

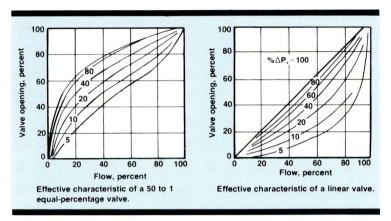

Effective characteristic of a 50 to 1 equal-percentage valve.

Effective characteristic of a linear valve.

Fig. 6-8. The Effect of Percent of Dynamic Drop across the Control Valve

6-8. Power Failure

It is inevitable that power will fail occasionally in a process operating unit and this produces a power failure for the final control element. Whether the final control element is pneumatic, hydraulic, or electric will change some of the individual aspects of the emergency.

In general, there are three possible patterns that might be encountered in power failure:

A. The final control element may fail open, forward, or upward (closed if a valve, forward if a rotary motor, and upward if a sliding stem). In general, this condition is referred to as "open."

B. The element may fail so as to hold the position which it last held. This is referred to as "last position."

C. The element may fail shut, reverse, or down. This is termed "closed."

In any given instance, it is possible that any one of these three situations may be desirable. Significant process analysis is required on the part of the practitioner to make specific decisions about the way in which the failure mode should be designed; it is an important part of the total process control strategy.

Exercises:

6-1. An equal-percentage control valve has 10 GPM flowing through it. When the air-to-valve signal increases 0.5 psi, flow through the valve increases 1.5 GPM. Much later, the base flow through the valve is operating at 20 GPM when the air-to-valve signal increases 0.5 psi; how much will flow increase through the valve?

6-2. Repeat Exercise 6-1 but assume the valve has linear trim.

6-3. A control valve flowing water has C_v of 20 and an assigned pressure drop of 10 psi. How many GPM will the valve flow?

6-4. A control valve is flowing air. It has a C_v of 32, a pressure drop assigned of 10 psi, an expansion factor of 0.8, and a specific weight of 3.103 lbs per ft^3. Calculate flow through the valve.

6-5. Given an equal-percentage valve installed with 10% of the dynamic pressure drop across the valve. Assume the valve operates about 50-70% open under normal conditions. Does the valve act more like an equal-percentage valve or more like a linear valve?

6-6. Consider several actual feedback loops either at home or at work, e.g., the heating system, the oven, the air compressor controller at your service station, etc. Decide for each loop how you would like the valve to "fail" in the event of power loss.

Unit 7:
Process Dynamics

UNIT 7

Process Dynamics

It is important that the student develop an overall appreciation of process dynamics. This unit introduces many of the general concepts of process dynamics and much of the associated terminology.

Learning Objectives — **When you have completed this unit, you should:**

A. **Know the general response characteristics of a first-order lag component which has been subjected to a step input.**

B. **Be able to determine graphically a time constant for a first-order lag system that has been driven by a step input.**

C. **Be able to identify a dead time on a process response curve.**

D. **Understand the effects of process lags and dead times on loop process dynamics.**

7-1. First-Order Lags

The *first-order lag* is the most common type of dynamic component encountered in process control. To study it, it is helpful to look at response curves when the component under study is subjected to a step input such as:

The advantage of using such a step input as a forcing function or driving force is that the input is at steady state for the time previous to the change, and then the input is instantaneously switched to a new steady state value. When the resulting output or response curve is inspected, you then observe the transition of the component as it passes from one steady state to a new steady state.

For a first-order lag component, the response to a step input is as illustrated in Fig. 7-1.

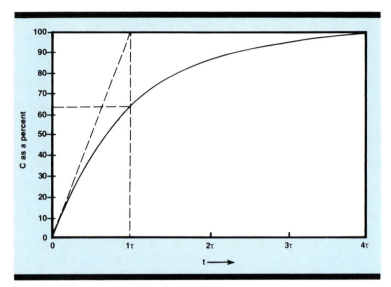

Fig. 7-1. The Response of a First-Order Lag Component to a Step Input

The component whose response is shown in Fig. 7-1 is called a first-order lag because the output lags behind the input and the differential equation which describes the underlying mathematics is a linear first-order differential equation.* This situation is the most common type of dynamic component encountered in practice. In addition to being referred to as first-order lag, these components are often referred to as *linear lags* or *exponential transfer lags*. These components are characterized by the capacity to store material or energy and the dynamic shape of these response curves is described by a *time constant*. The time constant τ is meaningful both in a physical

*The actual equation is:
$$\tau \frac{dc}{dt} + c = Kr$$

where c = output, r = input, K = gain, τ = time constant. In block diagram form, this is:

$$r \longrightarrow \boxed{\frac{K}{1 + \tau p}} \xrightarrow{c}$$

where p is the operator d / dt.

sense and in a mathematical sense. In a mathematical sense, it represents, at any moment, the future time necessary to experience 63.2% of the change remaining to occur. It assumes a step input has been introduced to the component under study. This is seen in Fig. 7-1.

If you look closely at the first-order response curve in Fig. 7-1, you also see that the response is always falling off, i.e., the rate of response is a maximum in the beginning and is continuously decreasing from that time onward. If the system were to continue to change at its maximum response rate (which of course occurs at the origin), it would reach its final value (or 100%) in 1 time constant.

Table 7-1 gives the numerical value of the changes that take place in a first-order lag response to a step input. In the first time constant of elapsed time, 63.2% of the total response will occur. Within the next time constant, 63.2% of the remaining 36.8% of the change will take place, etc. Theoretically, of course, the response never reaches 100%, but it does approach it asymptotically.

TABLE 7-1. RESPONSE OF A FIRST-ORDER LAG TO A STEP CHANGE

Elapsed Time	Percent of Total Response	Response Remaining	63.2 Percent of Response Remaining
1τ	63.2	36.8	23.2
2τ	86.4	13.6	8.6
3τ	95.0	5.0	3.16
4τ	98.16	1.84	1.16
5τ	99.32	0.68	0.429

The time constant gives specific dynamic insight into the rapidity with which the system or component responds; it is a specific measure of the rapidity of response. Fig 7-2 shows six different response curves, all for first-order lags, but each of the six has a different time constant. As you can see, as the time constant gets larger, the response gets slower.

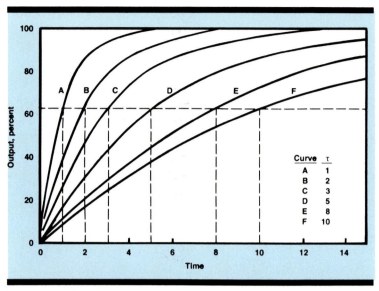

Fig. 7-2. Some First-Order Lag Responses to a Step Input

7-2. Time Constants

Frequent use is made of the time constant concept. It is an idea that is not entirely new to the technical person, and it has a good basis for understanding in simple electrical circuits. Shown below is a simple electrical RC network:

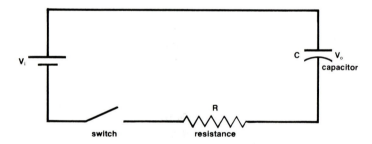

The basic problem is to derive the equation to describe the charging of this capacitor when a voltage V_i is suddenly applied to the circuit (by closing the switch). V_o is the voltage across the capacitor and it turns out that the buildup of V_o is given by a linear first-order lag.* The product RC, or resistance times

*The actual equation is:
$$RC \frac{dV_o}{dt} + V_o = V_i$$

capacitance, is the natural time constant for this simple network. If one checks the units involved, the product of RC has the units of time.

The product of resistance time capacitance is the system time constant for the simple network, but this physical analogy can also be used to gain insight into time constants of systems other than electrical networks. To do so, it is advantageous to use the reciprocal of resistance, which is conductance, and for many physical systems, time constants may be viewed more easily as capacitance divided by conductance:

$$\text{time constant} = (\text{resistance})\,(\text{capacitance}) = \frac{\text{capacitance}}{\text{conductance}}$$

Table 7-2 gives a broader view of the time constant analogy between various physical systems. In an earlier example (in Unit 4), you observed the first-order lag response of a bare bulb thermometer immersed in an agitated bath (a step input). In this case, the capacitance for the bulb is the mass of bulb material (which must change in temperature) times its weighted average heat capacity. The conductance is the heat transfer coefficient, between the bath liquid and the bulb, times the area through which heat transfer takes place. If this capacitance is divided by this conductance, you find the natural time constant for the bulb expressed in units of time.

It is important that you begin to get a general idea of the physical meaning of time constants. With practice, you can inspect an individual component or process and achieve some appreciation of its capacity to store material or energy, as well as for the conductance of material or energy into or through the component or process. In this way, you achieve specific insight into the time constants involved and, therefore, the process dynamics of an individual component or process are better understood. Such understanding will materially assist you in process control applications.

TABLE 7-2. TIME CONSTANT ANALOGY BETWEEN BASIC PHYSICAL SYSTEMS

Variable	Electrical	Liquid Level	Thermal	Pressure
Quantity	Coulomb	Cubic foot	Btu	Cubic foot
Potential or force	Volt	Foot	Degree	$\dfrac{\text{Pounds}}{\text{Square inch}}$
Flow rate	$\dfrac{\text{Coulombs}}{\text{Second}}$ = Amperes	$\dfrac{\text{Cubic feet}}{\text{Minute}}$	$\dfrac{\text{Btu}}{\text{Minute}}$	$\dfrac{\text{Cubic feet}}{\text{Minute}}$
Resistance	$\dfrac{\text{Volts}}{\text{Coulombs per second}}$ = ohms	$\dfrac{\text{Feet}}{\text{Cubic feet per minute}}$	$\dfrac{\text{Degrees}}{\text{Btu per minute}}$	$\dfrac{\text{Pounds per square inch}}{\text{Cubic feet per minute}}$
Capacitance	$\dfrac{\text{Coulombs}}{\text{Volts}}$ = farads	$\dfrac{\text{Cubic feet}}{\text{Foot}}$	$\dfrac{\text{Btu}}{\text{Degree}}$	$\dfrac{\text{Cubic feet}}{\text{Pounds per square inch}}$
Time	Seconds	Minutes	Minutes	Minutes

7-3. Higher-Order Lags

Of course, not all process dynamics can be neatly characterized by first-order lags. Often a situation is encountered in which a step input will produce a response curve such as shown below:

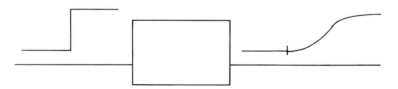

Note that the maximum rate of change for the output response curve does not occur at the origin but at some later time. This implies that a higher-order system is involved and a first-order lag is inadequate to characterize the system.

Higher-order systems can be the results of several different situations:

A. Several first-order lag processes may be encountered in series.

B. The installed feedback controller may introduce a characteristic differential equation which, when considered in series with the other system components, makes the overall description of the system higher ordered.

C. Mechanical or fluid components of the system may be subject to accelerations, i.e., to inertial effects (usually this is a minor possibility).

D. The process may be a *distributed* process which gives a response curve that can be described only by higher-ordered differential equations or by partial differential equations.

When first-order lags are encountered in series, the way in which they are connected in series will have a lot to do with the resulting process dynamics. They may be connected so as to produce *independent stages* or *interacting stages;* these are illustrated in Fig. 7-3. For example, both of these tanks taken separately would give first-order lag responses to a change in inlet flow rate. But the manner in which they are connected causes their combined response to differ between case (a) and case (b). In the (a) case the height of liquid in the second tank does not influence the flow of liquid out of the first tank, but in the (b) case, it is clear that the height of liquid in the second tank will influence the rate at which liquid flows out of the first tank. When stages are interacting with one another, the resulting process response will always be more sluggish than expected from simple first-order lags connected as independent stages. This case is of special importance in thermal systems because they are usually sluggish to begin with.

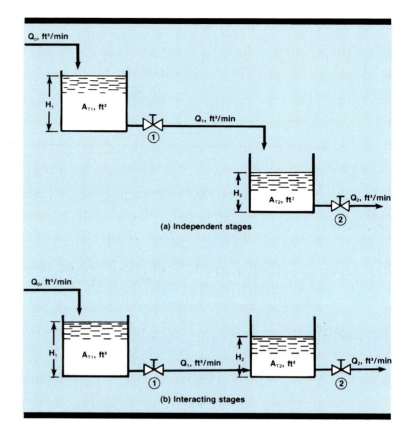

Fig. 7-3 Independent and Interacting Stages

Higher-ordered response, where the maximum change in process response rate does not occur at the origin, is quite common. Shown in Fig. 7-4 are some typical effects of thermowells on thermometer bulb response. Curve A is a plain bare bulb with first-order lag dynamics. As the bulb is provided with more protection, the response gets more and more sluggish, is higher ordered, and exhibits the S-shaped response curve of higher-ordered systems. In these cases there is, of course, much more lag involved and the process becomes more difficult to control.

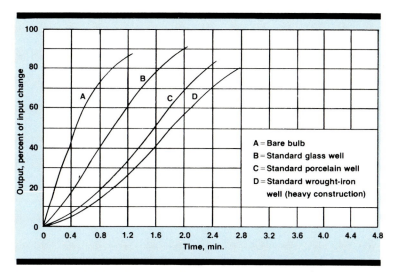

Fig. 7-4. Some Effects of Thermowells

7-4. Dead Time

In process dynamics you often encounter a process response curve in which there is a *dead time* before there is any dynamic response whatsoever. This might appear:

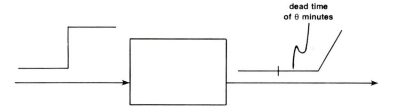

Such dead times are some of the most difficult situations to control. The difficulty stems from the fact that during this time period (the dead time) there is absolutely no response whatsoever, and therefore there is no information available to initiate corrective action.

For some specific insights into how a dead time might
occur in practice, refer to Fig. 7-5 which illustrates
simple feedback control of a system in which steam is
injected into a water tank to produce hot water. There
is a thermobulb in the outlet stream to measure the
temperature of the exit hot water. The control system
increases or decreases the manipulated variable, the
steam flow, to produce the outlet water at the desired
temperature.

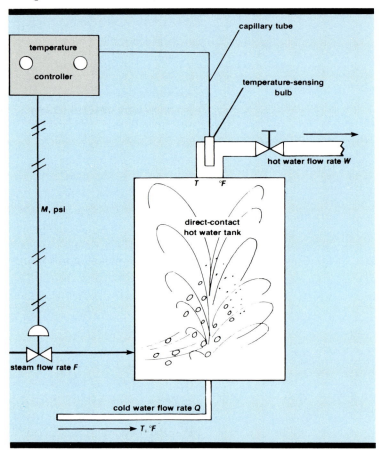

Fig. 7-5. Example System

The difficult question arises when one begins to
debate where this temperature bulb should be located,
i.e., where should the feedback sensor be installed? It
is tempting to say that it should be installed further
down the outlet pipe to be closer to the location where
the water actually will be used. This rationale sounds
appealing because it is at the point of usage that the
temperature of the water is significant, i.e., it is of

little importance what the water temperature might be in the tank itself or at the tank exit. This type of reasoning, while appealing, is disastrous. As the temperature bulb is moved further and further down the outlet pipe, a dead time is actually installed in the feedback loop before the sensor. In this case, dead time is equal to the *distance* down the pipe to the point at which the bulb is installed, divided by the *velocity* of the water inside the pipe. As a result of such a dead time encountered in the feedback loop, overall loop control deteriorates, and it may reach a point where it is impossible to achieve any feedback control at all.

Often dead times are referred to by other names such as *distance-velocity lags* or *transportation lags* or *time delays*. These other descriptive names can be appreciated very specifically in terms of the example shown in Fig. 7-5.

Quite often, dead times are unavoidable in actual practice, e.g., in a rotating kiln, in a plug flow reactor, or in similar situations. However, there are other cases where lack of insight and lack of forethought on the part of the designer cause the sensor to be installed in an inappropriate location and, for no good reason whatsoever, a dead time is inadvertently placed into the feedback control loop. Such mistakes cause serious operating problems and are a severe indictment of the designer. Sometimes such problems occur as a result of the type of reasoning discussed for the hot water tank; sometimes such problems are a result of a desire to place a sensor in a spot where maintenance might be achieved more easily, etc. Whatever the reason, dead times should be avoided wherever possible. They are easily the most difficult dynamic elements to control.

Dead times can occur in combination with nondynamic components:

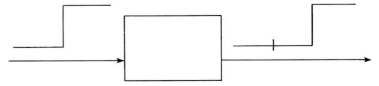

or, they may occur with first-order lags:

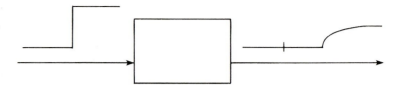

or they may be found in combination with higher-ordered lags:

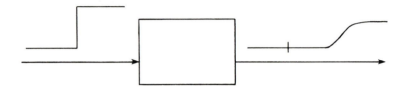

Whenever they are encountered, you should be able to simply pick them off the process response curve by a simple measure of the amount of time that transpires before any output response occurs after an input to the system.

7-5. Closed-Loop Response vs. Open-Loop Response

In the examples previously presented, attention was devoted to the open-loop response curve, i.e., the implication being that the feedback loop was not closed and that you were looking at the open-loop behavior of individual components or systems. The effective dynamic behavior of a component in a feedback loop will be much different when that loop is closed. This fact cannot be overemphasized.

One generalization needs to be made explicity clear. When a loop is closed around a dynamic component, generally it will become less stable, i.e., it will respond faster and more dynamically. This is a fact which all process operators realize very quickly. When they encounter dynamic control problems with a feedback loop, they immediately place the controller on "manual," i.e., they break the feedback loop and operate the control valve directly. In process control applications it is rare to encounter any open-loop oscillations, but if you take a system with several dynamic components, place these components together in a loop, and then close the loop, it is quite possible for the system to have an output response curve which exhibits oscillations. To discuss this more fully, you must become familiar with some of the descriptive terms used in characterizing systems that exhibit oscillation. Such descriptive terms are illustrated in Fig. 7-6.

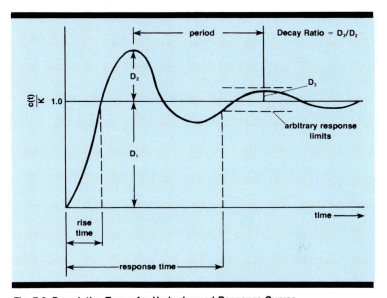

Fig. 7-6. Descriptive Terms for Underdamped Response Curves

To illustrate the effects of closing a process control loop, inspect the two response curves shown in Fig. 7-7. For illustration, assume two first-order lags in series. The overall gain of the two first-order lags is five and the two first-order lags have time constants equal to one another and, for illustrative purposes, equal to one. The open-loop response is shown. When the loop is closed and subjected to the same step input, the response curve is as shown in Fig. 7-7. Notice the dramatic change. This illustrates the need to make significant differentiations between open-loop and closed-loop response considerations.

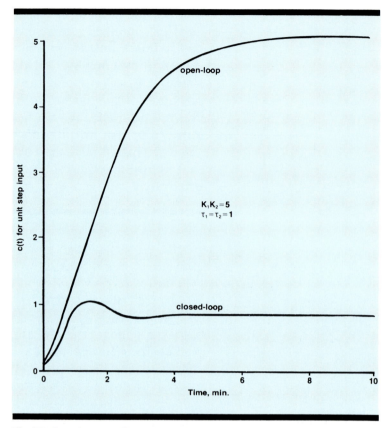

Fig. 7-7. Open-Loop vs. Closed-Loop Response

The point has been·made *that a loop is less stable when it is closed*. This generalization can be carried a step further. If you increase the gain on a loop that has been closed, the loop progressively becomes less and less stable, i.e., it exhibits more and more oscillatory behavior. In Fig. 7-7, if the gain on the loop had been higher, when the loop was closed the oscillation shown in the closed-loop response curve would have been larger. As you increase the gain higher and higher, you speed up the response of the loop. This is desirable because it speeds the controlled variable's transition due to setpoint changes and gives quicker response to disturbances. But as the gain is increased higher and higher, the loop becomes less and less stable and, therefore, there must be a trade off between the two phenomena. In effect, the *optimum gain* is the one that gives the proper trade off between speed of response and stability for the loop.

7-6. Some Generalizations

Some significant generalizations about process dynamics are especially important.

In the last section, it was demonstrated that when loops are closed, they become less stable, i.e., they show more significant dynamic changes, and that as the gain of a loop increases, it becomes less stable—but at the same time it responds faster.

In general, the more lag—the more time constants, and the more dead times—encountered around a process loop, the worse the control problem. The particular order or sequence in which the time constants or dead times are encountered is not especially significant. It does not make any difference what particular piece of hardware is involved, i.e., it does not make any difference whether a particular time constant is associated with the valve, with the sensor, or with the process itself. No matter where it occurs, it has the same deleterious effect on the overall control of the system.

These generalizations concerning process time constants include the controller. It was seen in Unit 5 that the addition of the reset mode to the controller also introduces some lag in the controller, and as a result, the loop is less stable when reset is added. The desirability of eliminating offset is often justification for reset, but you must appreciate the penalties involved in its usage.

There are cases in which the process may be so rapid that it, in effect, needs to be filtered or slowed down in order to increase controllability. There is one particularly important case of this—liquid flow control loops. In these, the process itself has almost instantaneous response and, as a result, there will be a lot of high-frequency noise transmitted around the loop. This is undesirable, and to counteract this, the reset mode is added to the flow controller in order to provide some lag or some filtering to the loop dynamics. In effect, lag is introduced via the reset mode in the controller. Of course, such an addition inadvertently eliminates offset, but the principal need is to smooth out some of the noise being transmitted around the loop.

The point is made that the more process lag, i.e., the more time constants and dead times around a loop, the worse the situation (except as noted in the previous paragraph). In addition, the more this process lag is distributed around the loop, the worse the control case. Saying this differently, the more the process lag is concentrated in a specific component, the better the control situation. For example, a first-order lag with a time constant of ten minutes would not be too difficult to control. If, however, this same process lag was broken into two time constants of five minutes each, the control would be more difficult. If it was broken into ten time constants of one minute each, it would be even worse. As a matter of fact, if it was broken into a large number, e.g., two hundred time constants of a few seconds each—but whose total was ten minutes—then the behavior would approach that of a ten minute dead time (the most difficult case of all).

Little can be done about process dynamics in many situations. The dynamic behavior of the process usually is established by the function it serves. It is expected that the practitioner will design and install a transmission system that operates sufficiently fast so that it does not contribute significant lag to the overall process control loop. This leaves the control valve and the sensor where dynamic choices are often made—sometimes without thinking. The dynamic behavior of these two components is especially important, and many control loops are made unnecessarily difficult to operate simply because the practitioner installs the sensor in such a way that it exhibits very slow and sluggish behavior or he installs the control valve in such a way that its lag is a significant contributing factor to the poor performance of the loop. The goal of the practitioner should be to have good appreciation of the dynamic behavior of all the hardware he designs and installs.

Exercises:

7-1. Estimate the time constants for the three components whose step response curves are shown below:

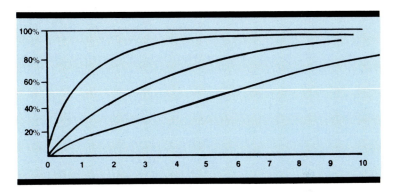

7-2. What can you say about the dynamics of a first-order lag type component that has a time constant of zero? Of infinity?

7-3. A stirred tank has a hold-up of 28 ft³ and a flow rate through it of 6 ft³/min. What is its time constant?

7-4. Consider the bare thermometer bulb of Fig. 4-1. The bulb has a mass of 0.26 lbs; a heat capacity of 0.6 BTU per lb per °F; heat transfer into the bulb occurs at a rate of 3.7 BTU per ft² per °F per min; heat transfer occurs through a bulb surface area of 0.16 ft². What is the bulb's time constant?

7-5. How much dead time is in the component whose step response is shown below:

7-6. Consider Fig. 7-5. If the bulb were installed 16 ft down the exit pipe and the flow velocity in the pipe were 21 ft per minute, how much dead time would there be in the feedback loop?

7-7. *Explain what happens in a feedback loop as you increase the controller gain higher and higher.*

7-8. *Explain what happens in a feedback loop as you introduce more and more lag, e.g., dead time, into the loop.*

Unit 8:
Tuning Control Systems

UNIT 8

Tuning Control Systems

A feedback control system is of little value if it is improperly tuned; the analogy to an improperly tuned automobile is instructive. It is important to have an understanding of how the controller in a feedback control system should be tuned.

Learning Objectives — **When you have completed this unit, you should:**

    A. Have developed insight into the fundamental concepts of tuning feedback controllers.

    B. Be able to calculate the tuning parameters for a feedback controller using the Ziegler-Nichols ultimate method.

    C. Be able to calculate the tuning parameters for a feedback controller using the process reaction curve method.

8-1. What Is Good Control?

The need in tuning a controller is to determine the optimum values of the controller gain K_c (or proportional band PB), the reset time T_i (or the reset rate as repeats per minute), and the derivative time T_d. The adjustment of these tuning parameters on feedback controllers is one of the least understood and most poorly practiced—yet extremely important—aspects of automatic control theory.

The first problem encountered in tuning controllers is to determine what *good* control is and, as might be expected, it does differ from one process to the next. The most common criterion employed is to adjust the controller so that the system's response curve has an amplitude ratio or *decay ratio* of one-quarter. A decay ratio of one-quarter means that the ratio of the overshoot of the first peak in the process response curve to the overshoot of the second peak is four to one. This is illustrated in Fig. 8-1.

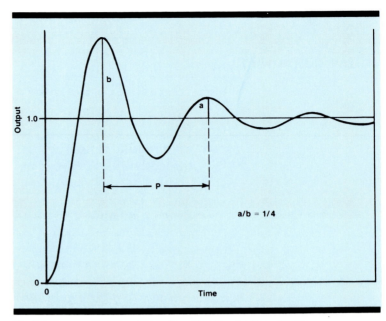

Fig. 8-1. Process Response Curve for a One-Quarter Decay Ratio

Basically, there is no direct mathematical justification for requiring a decay ratio of one-quarter, but it represents a compromise between a rapid initial response and a fast line-out time. In many cases, this criterion is not sufficient to specify a unique combination of controller settings, i.e., in two-mode or three-mode controllers there are an infinite number of settings which will yield a decay ratio of one-quarter, each with a different period. This illustrates the problem of defining what constitutes good control.

In some cases, it is important to tune the system so that there is no overshoot; in other cases, a slow and smooth response is desired; some cases warrant fast response and significant oscillations are no problem, etc. The point is—*you* must determine what control is good for each specific loop.

8-2. The Tuning Concept

The feedback controller is only one piece of hardware in the entire loop; there are many other hardware items connected to form the balance of the loop. For the purposes of adjusting the feedback controller, it is convenient and sufficient to view everything else within the feedback loop as being "one big lump." Actually, this is the way in which the feedback controller sees the balance of the loop. This is illustrated in Fig. 8-2.

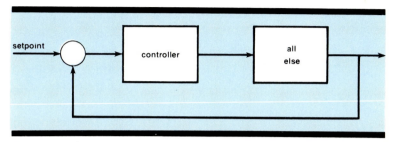

Fig. 8-2. The Balance of the Loop as "Seen" by the Feedback Controller

There is one parameter to adjust in a single-mode controller, e.g., the gain K_c in a proportional-only instrument. There are two parameters to adjust in a two-mode controller, e.g., the proportional gain K_c and the reset time T_i in a PI controller. There are three parameters to adjust in a three-mode controller, e.g., the controller gain K_c for the proportional mode, the reset time T_i for the integral mode, and the T_d for the derivative mode.* In adjusting the controller, the gains around the loop will tend to dictate what should be the optimum gain in the controller. Similarly, the time constants and dead times that characterize the lag dynamics of the *all else* will tend to dictate what will be the optimum value of the reset time and what the derivative time in the controller should be. Stating this differently, before you can calculate or select the best values for the tuning parameters in the controller, you must get some quantitative information about the overall gain and the process lags that are present in the balance of the feedback loop. This illustrates quite clearly why controllers cannot be preset at the factory but, instead, must be individually tuned for individual loops.

It is also helpful to establish a mathematical model of a process control loop and appreciate the role of the controller as a mathematical equation within that

*Murrill's Law states that the difficulty of tuning a controller increases with the square of the number of modes present. Humor aside, it tends to be correct!

model. As you saw earlier, each of the individual blocks around the feedback loop represents an algebraic or differential equation, i.e., it represents a mathematical statement for that particular piece of hardware. Since all of these blocks are coupled together, all of these equations represent a simultaneous set of mathematical equations. If you are able to determine what you think good control is, then you can, in effect, specify the overall solution to this simultaneous set of equations. Basically, the tuning of the controller represents the adjustment of the individual parameters in the equation which represents the controller. As you adjust these parameters for the controller equation, you modify the solution for the simultaneous set, i.e., you change the response of the overall system toward a point that represents good response or good control.

8-3. Closed-Loop Tuning Methods

Techniques for adjusting controllers may be classified as either open-loop or closed-loop methods. One of the first methods proposed for tuning feedback controllers was the *ultimate* method proposed by Ziegler and Nichols in 1942. The term *ultimate* was attached to this method bcause its use required the determination of the *ultimate gain* (sensitivity) and *ultimate period* for the loop. The ultimate gain is the maximum allowable value of gain (for a controller with only a proportional mode in operation) for which the closed-loop system is stable.

For any feedback control system, if the loop is closed (if the controller is on automatic), you can increase the controller gain and, as you do so, the loop will tend to oscillate more and more. As you continue to increase the gain further, you will observe continuous cycling or continuous oscillation in the controlled variable. This is the maximum gain at which the system may be operated before it becomes unstable; this is the ultimate gain. The period of these sustained oscillations is the ultimate period. If you increase the gain further, the system will become unstable. These general situations are illustrated in Fig. 8-3.

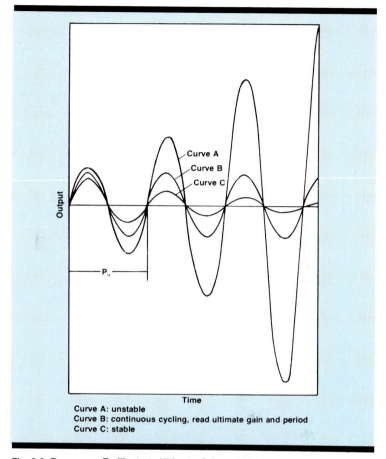

Curve A: unstable
Curve B: continuous cycling, read ultimate gain and period
Curve C: stable

Fig. 8-3. Responses To Illustrate Ultimate Gain and Ultimate Period

To determine the ultimate gain and the ultimate period, take the following steps:

1. Tune out all the reset and derivative action from the controller, leaving only the proportional mode, i.e., set T_i equal to infinity and T_d equal to zero (or as close to these values as is possible on the controller).

2. Maintain the controller on automatic, i.e., leave the loop closed.

3. With the gain of the proportional mode of the controller at some arbitrary value, impose an upset on the process and observe the response. One easy method for imposing the upset is to

move the setpoint for a few seconds and then
return it to its original value.

4. If the response curve from Step 3 does not damp
    out (as in Curve A in Fig. 8-3), the gain is too high
    (proportional band setting too low). The gain
    should be decreased (proportional band setting
    should be increased) and Step 3 repeated.

5. If the response curve in Step 3 damps out (as in
    Curve C in Fig. 8-3), the gain is too low
    (proportional band is too high), the gain should be
    increased (proportional band setting should be
    decreased) and Step 3 repeated.

6. When a response curve similar to Curve B in Fig.
    8-3 is obtained, the values of the ultimate gain (or
    ultimate proportional band) setting and the
    ultimate period of the associated response curve
    are noted. This ultimate gain at which the
    sustained oscillations are encountered is the
    ultimate sensitivity S_u and the ultimate period is
    P_u.

The ultimate gain and the ultimate period are then
used to calculate controller settings. Ziegler and
Nichols correlated in the case of proportional
controllers that a value of operating gain equal to
one-half of the ultimate gain would often give a decay
ratio of one-quarter, and they therefore proposed a
tuning rule-of-thumb for a proportional controller:

$$K_c = 0.5 \ S_u \tag{8-1}$$

By similar reasoning and testing, the following
equations were found to represent good
rules-of-thumb for controller settings for more
complex controllers.

Proportional-plus-reset:

$$K_c = 0.45 \ S_u \tag{8-2}$$
$$T_i \ = P_u/ \ 1.2 \tag{8-3}$$

Proportional-plus-derivative:

$$K_c = 0.6 \ S_u \qquad\qquad (8\text{-}4)$$
$$T_d = P_u / 8 \qquad\qquad (8\text{-}5)$$

Proportional–plus–reset–plus–derivative:

$$K_c = 0.6 \ S_u \qquad\qquad (8\text{-}6)$$
$$T_i = 0.5 \ P_u \qquad\qquad (8\text{-}7)$$
$$T_d = P_u / 8 \qquad\qquad (8\text{-}8)$$

Again, it should be noted that the above equations are empirical and exceptions are inherent. They generally are intended to achieve a decay ratio of one-quarter, i.e., this is the inherent definition of good control.

Example 8-1: For a temperature control system whose ultimate sensitivity S_u is 0.4 psi/°C and ultimate period is two minutes, determine settings for various controllers using the Ziegler-Nichols ultimate method.

Proportional:

$$K_c = 0.5 \ S_u = 0.2 \ \text{psi/°C}$$

Proportional-plus-reset:

$$K_c = 0.45 \ S_u = 0.18 \ \text{psi/°C}$$
$$T_i = P_u/1.2 = 1.67 \ \text{min}$$

Proportional-plus-reset-plus-derivative:

$$K_c = 0.6 \ S_u = 0.24 \ \text{psi/°C}$$
$$T_i = 0.5 \ P_u = 1.0 \ \text{min}$$
$$T_d = P_u/8 = 0.25 \ \text{min}$$

A slight modification of the ultimate method has also been proposed by Harriott. For many processes it is inconvenient to allow sustained oscillations and the ultimate method cannot be used. In Harriott's modification of the ultimate method, the gain (proportional control only) is adjusted until a step response curve with the decay ratio of one-quarter is obtained. It is necessary to note only the period P of this response, and with this value P the reset and derivative modes are set as:

$$T_i = P/6 \tag{8-9}$$
$$T_d = P/1.5 \tag{8-10}$$

After setting these modes, the controller gain is again adjusted until a response curve with a decay ratio of one-quarter is obtained. This method usually requires about the same amount of work as the ultimate method, since often it is necessary to experimentally adjust the value of the gain determined from the ultimate method itself to obtain a decay ratio of one-quarter.

Example 8-2. Suppose a process presently controlled by a proportional controller has a response whose period is three minutes when the decay ratio is one-quarter. If reset and rate action are added to the controller, what settings are recommended?

$$T_i = P/6 = 3 \text{ min}/6 = 0.5 \text{ min}$$
$$T_d = P/1.5 = 3 \text{ min}/1.5 = 2 \text{ min}$$

There are other closed-loop methods and many of them are conceptually equivalent or similar to the two methods presented.

8-4. The Process Reaction Curve

Generally speaking, open-loop methods require only that a single upset be imposed on the process. These methods give more precise data about the dynamics of a feedback control system and usually they give slightly better tuning results—though the variation in satisfaction from loop to loop can be quite significant.

The process reaction curve is basically the *reaction* of the process to step change in its input signal. The process reaction curve is the reaction of the *all else* viewed in Fig. 8-2. It is important to get a complete picture of exactly what this process reaction curve might represent in a feedback loop and this is shown in some detail in Fig. 8-4. Note that the air-to-valve signal can be used to introduce the step change to the overall process and the dedicated trend recorder on the feedback loop may be used to record the process reaction curve. In general, a process reaction curve can be determined as follows:

1. Let the system come to steady state.

2. Place the controller on manual operation, i.e., remove it from automatic operation.

3. Manually set the air-to-valve signal at the value at which it was operating automatically.

4. Allow the system to reach steady state.

5. With the controller still in manual operation, impose a step change in the air-to-valve signal.

6. Record the response of the controlled variable. (Although the response is often being recorded by a dedicated trend recorder for the loop, it is often desirable to have a supplementary recorder available with a faster chart drive. This arises from the fact that dedicated trend recorders have very slow chart drives and small charts.)

7. Return the air-to-valve signal to its previous value and return the controller to automatic operation. Once this has been done, the recorded process reaction curve may be used to give significant information about the overall dynamics and characterizing parameters for the *all else* of the process loop. This information may be used to calculate needed tuning parameters of the feedback controller.

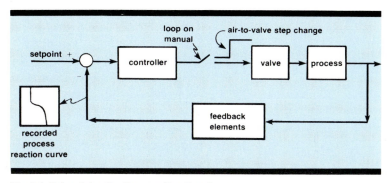

Fig. 8-4. Determining the Process Reaction Curve

8-5. A Simple Open-Loop Method

One of the earliest methods using the process reaction curve for tuning controllers was also proposed by Ziegler and Nichols (in the same article that presented the ultimate method). To use this process reaction curve method, only the parameters R_r and L_r must be determined. An example determination of these parameters is illustrated in Fig. 8-5 for a specific temperature control loop:

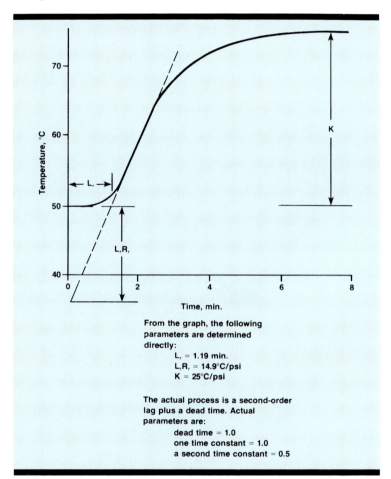

From the graph, the following parameters are determined directly:

L_r = 1.19 min.
L_rR_r = 14.9°C/psi
K = 25°C/psi

The actual process is a second-order lag plus a dead time. Actual parameters are:

dead time = 1.0
one time constant = 1.0
a second time constant = 0.5

Fig. 8-5. Process Reaction Curve for an Example Feedback Control Loop

To obtain process information parameters in the process reaction curve method, a tangent is drawn to the process reaction curve at its point of maximum slope. This slope is R_r, the process reaction rate. Where this tangent line intersects the original base line gives an indication of L_r, the process lag. L_r is really a measure of equivalent dead time for the process. If this tangent drawn at the point of maximum slope is extrapolated to a vertical axis drawn at the time when the step was imposed, then the amount by which this is below the horizontal base line will represent the product L_rR_r. Using these parameters, Ziegler and Nichols proposed a series of rules-of-thumb that can be used to calculate appropriate controller settings:

Proportional only:

$$K_c = 1/L_rR_r \qquad\qquad\qquad (8\text{-}11)$$

Proportional-plus-reset:

$$K_c = 0.9/L_rR_r \qquad\qquad\qquad (8\text{-}12)$$
$$T_i = 3.33\ L_r \qquad\qquad\qquad (8\text{-}13)$$

Proportional-plus-reset-plus-rate:

$$K_c = 1.2/L_rR_r \qquad\qquad\qquad (8\text{-}14)$$
$$T_i = 2.0\ L_r \qquad\qquad\qquad (8\text{-}15)$$
$$T_d = 0.5\ L_r \qquad\qquad\qquad (8\text{-}16)$$

Ziegler and Nichols indicated these rules-of-thumb should give a decay ratio of one-quarter, i.e., this is the inherent definition of good control.

In summary, to use this particular process reaction curve method, it is necessary to obtain a specific process reaction curve, and then, using the L_r and the R_r graphically determined from the process reaction curve, it is possible to calculate the needed tuning parameters.

Example 8-3: In Fig. 8-5 is a process reaction curve in which it is determined that

$$L_r = 1.19\ \text{min}$$
$$L_rR_r = 14.9°\text{C/psi}$$
$$K = 25°\text{C/psi}$$

Using Ziegler and Nichols' open-loop method,

calculate needed controller settings:

Proportional only:

$$K_c = 1/L_r R_r = 1/(14.9°C/psi) = 0.067 \text{ psi/°C}$$

Proportional-plus-reset:

$$K_c = 0.9/L_r = 0.9/(14.9°C/psi) = 0.06 \text{ psi/°C}$$
$$T_i = 3.33 \, L_r = (3.33) \, (1.19 \text{ min}) = 3.9 \text{ min}$$

Proportional-plus-reset-plus-rate:

$$K_c = 1.2/L_r R_r = 1.2/(14.9°C/psi) = 0.081 \text{ psi/°C}$$
$$T_i = 2.0 \, L_r = (2.0) \, (1.19 \text{ min}) = 2.38 \text{ min}$$
$$T_d = 0.5 \, L_r = (0.5) \, (1.19 \text{ min}) = 0.59 \text{ min}$$

8-6. Integral Methods

In addition to the types of process tuning methods already presented, there have been developed in recent years a large number of tuning techniques which are based on minimizing the values of various integral criteria. Usually, these try to minimize the value of an error function* over all time. Basically, integral criteria techniques are well suited for computer control applications and are recommended only for such installations. In such cases they can give very good tuning results.

8-7. The Need to Retune; Adaptive Tuning

Feedback control loops may be mathematically classified as *linear* or *nonlinear*. If a loop is linear it will simply double its output—via the exact-shape dynamic path—if you double its input. In a similar vein, if an input is a mirror image of a previous input, the output or response will be a mirror image. If a control loop is linear, it may be tuned once and it is tuned forever. But if a loop is nonlinear, its tuning will require continuous adjustment (or it will be poorly tuned). This phenomena is a major problem in keeping control systems well tuned.

Common criteria minimize the error squared, the absolute value of the error, or the absolute value of error times time.

It does not matter whether any one individual component of the control loop is linear or not. What matters is whether or not the *all else*—everything except the controller—is, in its aggregate form, linear or not. Sensors and transmission systems are designed by vendors as linear. When one selects the valve trim, it is important to select valve trim and assign pressure drop across the valve so that its performance (see Fig. 6-8), when coupled with the performance of the process, produces a linear combination.

One very exciting aspect of the advances being made in the tuning of industrial control systems has to do with the concept of self-tuning or *adaptive tuning*. This idea is based on the concept that one should be able to design and implement controllers capable of tuning themselves or of being tuned automatically. Such control systems and individual controllers do exist, but until recently they were so expensive that they were not practical for routine application. The design and installation of systems and controllers capable of adaptively tuning themselves, i.e., adapting themselves to the process situation manifest at a given moment, has been practical only for situations such as aircraft, special reactors, paper machines, etc. But microprocesses and digital capabilities have changed all this! Today the computational capabilities to do adaptive tuning are relatively cheap and there is a strong movement toward the implementation of such systems. Several different techniques for doing adaptive tuning are being used in industrial practice today; the concepts are not especially complex, but they are beyond the scope of this ILM.

8-8. Summary

There are literally scores of different tuning techniques available and they vary considerably in philosophy and implementation. Generally speaking, they vary in terms of their inherent definition of good control, whether they are open-loop or closed-loop, how complex they might be mathematically, etc. Quite often, they give widely varying results. It is virtually impossible to say that one technique is clearly superior to all others.

It is fair to say that much of the tuning of control systems is an art. It requires a great deal of practice and experience to develop a sensitive and professional *feel* for the tuning of systems. Recognizing the many difficulties associated with tuning control systems, ISA has developed a specific ILM on tuning industrial control systems. The practitioner interested in developing more detailed insight into this particularly important subject should refer to this separate ILM.

Exercises:

8-1. *How is it physically possible in two-mode and three-mode controllers (such as PI or PID) for there to be more than one set of controller settings that give a one-quarter decay ratio?*

8-2. *Why do not all liquid level control loops have the same tuning parameters, all flow control loops have the same tuning parameters, etc?*

8-3. *For a level control loop, whose ultimate gain is 0.3 psi per ft and whose ultimate period is three min, calculate the settings for various types of controllers using the ultimate method.*

8-4. *Inspect Fig. 8-5 and note that the gain K of the **all else** may be obtained. What are the units of this K? What are the units of the products of all the gain around a feedback control loop, i.e., the units of KK_c?*

8-5. *An experimental process reaction curve is run for a level control loop. From it, $L_r = 0.15$ min and $R_r = 0.6$ ft per psi per min. Calculate controller settings for various types of feedback controllers.*

8-6. *Convert the results of Exercise 8-5 for a controller using proportional band, reset rate, and derivative time.*

Unit 9:
Cascade, Ratio and Dead Time Control

UNIT 9

Cascade, Ratio and Dead Time Control

To this point attention has been focused on simple, single-loop feedback control. This unit will now expand horizons to several variations of this basic strategy.

Learning Objectives — **When you have completed this unit, you should:**

A. **Know the basic principles of cascade control.**

B. **Be able to explain the general guidelines for cascade controller mode selection and tuning.**

C. **Understand the basic concepts of ratio control.**

D. **Understand the basic principles of dead time control.**

9-1. Cascade Control

The general concept of cascade control is to nest one feedback loop inside another feedback loop. This is illustrated graphically in Fig. 9-1. In effect, you take the process being controlled and find some intermediate variable within the process to use as the controlled variable for the inner loop—take the process and *split* or divide the lag into two parts.

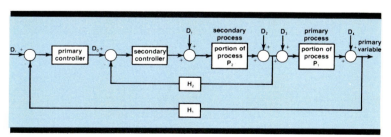

Fig. 9-1. The Concept of Cascade Control

Cascade control exhibits its real value when a very slow process is involved. When this happens, errors can exist for very long periods of time, and when disturbances enter the process, there may be a significant wait before any corrective action is initiated. Also, when corrective action is taken, you may have to wait a long time for results. Cascade control affords you the opportunity to find intermediate controlled variables and to take corrective action on disturbances more promptly.

The use of cascade control appears to involve significant additional hardware expenditures. As can be seen from the general layout, it requires an additional feedback controller, and it appears to involve an additional sensor and feedback transmission system. There is no need, of course, for an additional final control element such as a control valve. These general appearances are a bit misleading, however; usually hardware vendors supply both the primary and secondary controllers for the cascade arrangement within a single controller case, and the total cost for the two controllers is not twice the cost of a single controller. In addition, it is probably true that the intermediate or secondary controlled variable is of sufficient importance that often it is already one of the variables that is sensed and transmitted back to the central control room for indicating and/or recording purposes. If this is the case, there is no incremental cost associated with the secondary controlled variable's sensor and feedback transmission system. In general, the incremental hardware costs associated with implementing cascade control are not prohibitive. As a matter of fact, these costs are not usually a significant factor.

In general, cascade control has significant advantages to the user and is one of the most underutilized feedback control techniques. Most plants could increase the usage of cascade control to significant advantage.

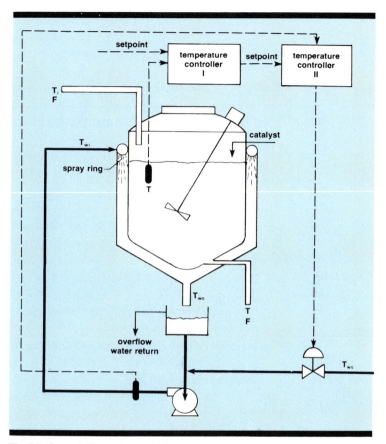

Fig. 9-2. Cascade Control on a Jacketed Kettle

An example will further our understanding; consider
Fig. 9-2. If this kettle had a simple feedback control
loop without cascade provisions, the sensor on the
primary controlled variable (the reacting mass
temperature) would go to the feedback controller and
the output of the feedback controller would directly
adjust the manipulated variable (the make-up cooling
water supply). The jacketed kettle as presented has a
cascade arrangement with the primary controlled
variable still being the temperature of the reacting
mass. An intermediate controlled variable for the
secondary controller is the temperature of the
circulating cooling water for the jacket of the kettle. In
this case, as is standard in cascade control

arrangements, the manipulated variable for the primary or master controller is the setpoint for the secondary or slave controller.

In order to understand cascade control, it will be helpful to look at two potential disturbances for the example system. First, consider what happens when there is a change in inlet feed temperature to the jacketed kettle. A change in this variable T_i will change the basic temperature of the reacting mass. In a general sense, the reacting mass will change in temperature by a first-order lag type relationship; the time constant is the mass of the contents of the vessel times its heat capacity (capacitance) divided by the flow rate F of fluid into the vessel times its heat capacity (conductance). This time constant for the contents of the kettle probably will be quite large. As the reacting mass starts to change in temperature, it will be sensed by the primary sensor which has its own dynamics—at least a first-order lag—and this first-order lag is *interacting* with the time constant of the kettle contents. The primary controller receives the error caused by the change in temperature, and corrective action is taken by adjusting the setpoint on the secondary or slave controller. Then corrective action is taken by adjusting the manipulated variable, the make-up cooling water flow.

As an alternative upset or disturbance, consider what happens if the cooling water supply has a change in its temperature T_{ws}. When this happens, there soon will be a change in the temperature of the discharge of the cooling water circulating pump, and this change in temperature will be sensed by the sensor for the secondary controller. If there were no cascade control, this disturbance would not be *sensed* until the temperature change worked through the entire cooling water system and began to change the temperature of the reacting mass inside the vessel.

The advantages of the cascade arrangement for this type of upset are apparent.

Fig. 9-3 gives a general block diagram layout for this temperature control system and the two disturbances are shown entering the system. The improvement in handling disturbances in the cooling water temperature is obvious.

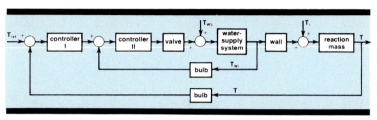

Fig. 9-3. Block Diagram for Jacketed Kettle Cascade Control System

9-2. Guiding Principles for Implementing Cascade Control

A tough question in implementing cascade control is how to find the most advantageous secondary controlled variable, i.e., to determine how the process can best be divided. In the selection of this intermediate point, quite often there is a large number of choices available to the designer. The overall strategy or goal should be to get as much of the process lag into the outer loop as possible while, at the same time, having as many of the disturbances as possible enter the inner loop.

Fig. 9-4 shows a general layout of a fired charge heater which is used to increase the temperature of a fluid charge passing through a fired furnace. The feedback control of this arrangement is shown in Fig. 9-4(a). Also shown in Fig. 9-4 are three different ways to establish a cascade control arrangement. In every case the primary controlled variable is the same, but in each case a different intermediate controlled variable has been selected. Clearly the question is which type of cascade control is best.

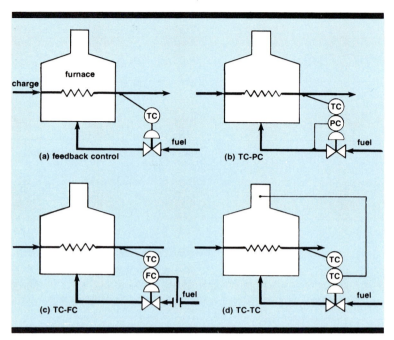

Fig. 9-4. Alternate Cascade Control Arrangements on a Fired Charge Heater

In order to determine the best cascade control arrangement, it is necessary to make a specific determination of what the most likely disturbances to the system are. It is helpful to make a list of these in order of decreasing importance. Once this has been done, the designer may review the various cascade control options available and determine which best meets the overall strategy outlined earlier, i.e., to have the inner loop as fast as possible while at the same time receiving the bulk of the important disturbances. (It is almost of secondary interest to this particular discussion, but in many petrochemical operations, the basic argument is whether the FC-TC or the TC-TC arrangement is preferred. There is much debate as to which is best.)

9-3. Selection of Cascade Controller Modes and Tuning

If both controllers of a cascade control system are three-mode controllers, here is a total of six tuning adjustments. It is doubtful if such a system could ever be tuned in an effective manner!* In selecting the modes to be included in both the primary and

_____
* *Remember Murrill's Law.*

secondary controllers of a cascade arrangement, the
burden of proof should be on proving why a control
mode should be added.

For the secondary (or inner or slave controller) it is
standard practice to include the proportional mode.
There is little need to include the reset mode for the
purposes of eliminating offset since the setpoint for
the inner controller will be reset continuously by the
outer or master controller. Occasionally, reset is added
to the inner loop controller if it is a liquid flow loop
because of the need to filter some of the transmission
of high-frequency noise around the loop. This use of
reset was discussed earlier.

For the outer loop, the controller should contain the
proportional mode and, if the loop is sufficiently
important to merit cascade control, it is probably true
that reset should be included to eliminate offset in the
outer loop.

The use of rate or derivative control in either loop
should be undertaken only if the loop has a very large
amount of lag.

The tuning of cascade controllers is the same as the
tuning of all feedback controllers, but the practitioner
must *work from the inside out*. The master controller
should be put on manual, i.e., the loop broken, and
then the inner loop can be tuned by whatever tuning
technique the practitioner finds most useful. Once the
inner loop is properly tuned, then the outer loop may
be tuned. In doing so, the outer loop "sees" the tuned
inner loop functioning as part of the total *process* or
the *all else* that is being controlled by the master
controller. If one follows this general *inside-first*
principle in tuning cascade controllers, no special
problems should be encountered.

9-4. Ratio Control

Another commonly encountered type of multiple-loop
feedback control system is *ratio control* (sometimes
called *fraction* control). When one looks at just the
hardware, ratio control quite often is confused with

cascade control because in ratio control, one loop adjusts another. But basically, the operation of ratio control is quite different.

Ratio control is often associated with process operations in which it is necessary to mix two or more streams together continuously to maintain a steady composition in the resulting mixture. A practical way to do this is to use a conventional flow controller on one stream and control the other stream with a ratio controller which maintains that stream flow at some preset ratio or fraction to the primary stream flow.

A ratio control system for regulating the composition of a feed stream to a reactor is shown in Fig. 9-5. The block diagram arrangement for this system is shown in Fig. 9-6. In this diagram, the subscript $_1$ refers to the airstream in Fig. 9-5 while the subscript $_2$ refers to the hydrocarbon stream. K is the ratio or fraction which is adjustable. The "flow fraction" element is actually just a multiplier with an externally adjusted gain.

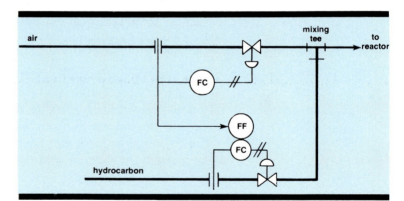

Fig. 9-5. A Conventional Ratio Control System

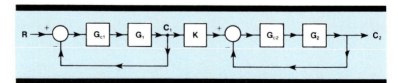

Fig. 9-6. Conventional Ratio Control System's Block Diagram

The design of a ratio control system poses no special problem because each of the loops is designed

individually and all of the general principles
presented earlier still apply.

It is also possible to implement a ratio control system
if the primary instrument is not a controller but a
transmitter. In such a situation, the setpoint of the
controller is set in direct relation to the magnitude of
the primary controlled variable. An example of this is
shown in Fig. 9-7 and the associated block diagram
for this type of ratio system is shown in Fig. 9-8.
Basically, the general principles are very similar to
other ratio control systems except one of the streams
is *uncontrolled* and the other is simply maintained in
ratio to it.

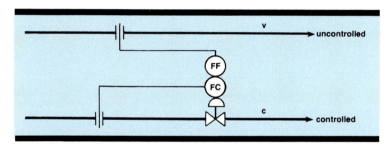

Fig. 9-7. Ratio Control with an Uncontrolled Stream

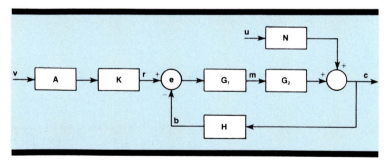

Fig. 9-8. Block Diagram for Ratio Control with an Uncontrolled Stream.

9-5. Dead Time Control

Dead time has been labeled by many as *the most
difficult element to control*. There are some special
feedback control arrangements suitable for the control
of processes in which large significant dead times are
present.

Fig. 9-9(a) shows a general layout for feedback control of a loop. In this case, assume there is a significant amount of dead time θ present in the loop. In conventional feedback control, as shown in Fig. 9-9(a), the variable fed back to the controller is the output of the dead time θ and this causes a difficult control problem.

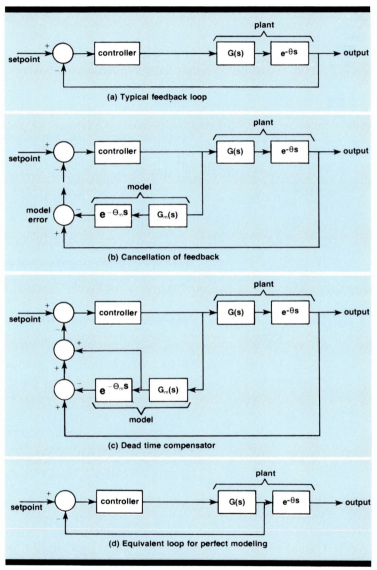

Fig. 9-9. Dead Time Control (s is Laplace variable notation)

If in some way this feedback variable could be placed (effectively) outside of the loop, i.e., if the dead time the controller *sees* could be moved outside the loop, the controller then could be tuned much more tightly and could provide decent control to the loop. This sounds good in practice but, of course, quite often the process and its associated dead time cannot be separated as distinctly and neatly as shown in Fig. 9-9(a). This does not prevent, however, modeling the total plant as though this were the case.

Fig. 9-9(b) shows a feedback system model that is operated in parallel with the actual plant itself. Θ_m is model dead time in this figure, the output of the controller is an input to both the model and to the actual plant. The output of the model is used to cancel the original feedback signal. If the model is perfect, the output of the summer (the model error) will be zero. Shown in Fig. 9-9(c), the output from the model plant before the dead time can be used as a *feedback* to the feedback controller. For the realistic case of imperfect modeling, the output from the lower summer is the modeling error, and it can be used to adjust the output of the model which is providing feedback to the controller. The general overall operation, with perfect modeling, is shown in Fig. 9-9(d).

For a complex control problem such as is illustrated in this example of dead time control, it is clear that more than conventional off-the-shelf hardware is required. Typically, the scheme illustrated in Fig. 9-9 can (and would) be implemented using digital hardware. It is presented here principally to illustrate the fact that special feedback arrangements can be devised to handle special problems.

Exercises:

9-1. *Refer to the jacketed kettle of Fig. 9-2. Are you better off or worse off with cascade control (as compared to plain feedback control) if the inlet feed temperature T_i changes?*

9-2. *Given the level controller LC below:*

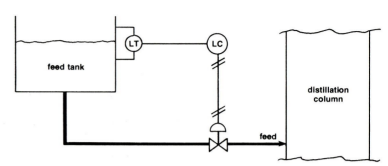

Make this a cascade arrangement by installing a liquid flow control loop on the column feed. What are the advantages?

9-3. *Given the composition controller CC on the reflux flow to a distillation column:*

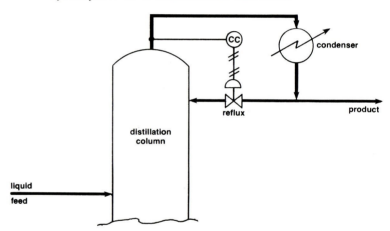

Add a liquid flow controller on the reflux as a cascade arrangement. Now add a level controller on the reflux condenser to govern product withdrawal.

9-4. *Draw a conventional ratio flow control system to blend continuously a 10 to 1 dry martini. Redo with an **uncontrolled** gin stream.*

9-5. *Diesel oil is injected into the suction of a fuel oil pump (the impeller does the mixing) to control viscosity on fuel oil:*

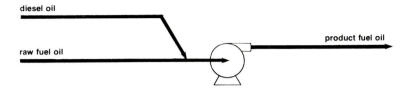

Design a ratio control system.

9-6. Install a viscosity meter as a sensor on the product fuel oil line of Exercise 9-5 and use it to adjust the ratio between the two streams.

9-7. Where should the sensor of Exercise 9-6 be located? What are the problems associated with moving it further downstream to the point where the product fuel oil is delivered to a customer?

Unit 10:
Feedforward and Multivariable Control

UNIT 10

Feedforward and Multivariable Control

All of the previous units have focused on feedback control; it is now desirable to describe some of the inadequacies of feedback control and to show how feedforward control can be used to advantage.

Learning Objectives — When you have completed this unit, you should:

A. **Understand the basic concepts of feedforward control.**

B. **Be able to outline the structure of the control equation contained in a feedforward controller.**

C. **Understand the problems associated with multivariable control and the solution approaches that are available to solve these problems.**

10-1. Feedforward Control

There are two process conditions which can make the overall effectiveness of feedback control quite unsatisfactory. One of these is the occurrence of disturbances of large magnitude and the other is the occurrence of large amounts of process lag. The question of importance of either occurence is defined in economic terms. In either case, the principal concern is the existence of errors that have significant economic consequences in overall process operations. Feedforward control can be used to deal with these disadvantages or inadequacies of feedback control, i.e., to deal with these errors.

In Unit 2, the overall general concept of feedforward control was introduced and it is now time to view this in terms of a single loop. Fig. 10-1 shows the possibilities of a single controlled variable for a process subject to a single significant disturbance. A manipulated variable is chosen and the selection of the manipulated variable is based on many of the same criteria typically used in feedback control.

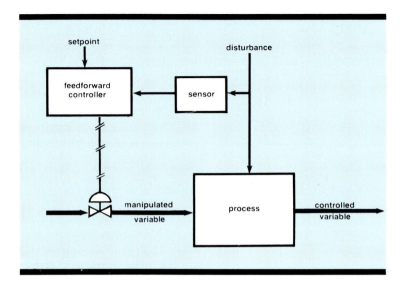

Fig. 10-1. Feedforward Control Loop

In feedforward control, however, a sensor is used to measure the disturbance as it enters the process and the sensor transmits this information to a feedforward controller. The feedforward controller determines the needed change in the manipulated variable so that when the effect of the disturbance is combined with the effect of the change in the manipulated variable, there will be no change in the controlled variable. This perfect compensation is a difficult goal to obtain. It is, however, the objective for which feedforward control is structured. Just as in feedback control, it is necessary to provide the feedforward controller with a setpoint or desired value of the controlled variable.

There are some significant difficulties in feedforward control. The structure of feedforward control assumes that the disturbances are known in advance, that the disturbances will have sensors associated with them, and that there will not be significant *unsensed* disturbances. There is also a tremendous escalation of theoretical know-how required in the feedforward controller's computation activities. In feedback control, relatively standard control algorithms (such as the P, PI, or PID controllers) were used, but in feedforward control, each controller equation or algorithm is specifically and uniquely designed for the one particular control application involved. In effect, the feedforward control computation involves determining exactly how much change in manipulated variable is required for a specific change in disturbance. To be able to make this computation accurately requires significant quantitative understanding of the process and its operation.

There is one other significant aspect of pure feedforward control as shown in Fig. 10-1 that merits consideration. In this case, there are no feedback phenomena whatsoever; if the controlled variable strays from its setpoint or desired value, the control system is unaware and takes no corrective action to eliminate the deviation. This makes pure feedforward control somewhat impractical and a rarity in typical process applications. The usual case is to combine feedforward control with feedback control.

It can be seen that feedforward control requires a significant increase in technical skills and capabilities. As a result, feedforward control of specific variables is limited to the most economically significant cases. In practical industrial applications, only very few cases are handled with feedforward control. These cases are, however, usually very significant, and the overall subject of feedforward control thus merits study.

10-2. A Feedforward Control Example

In Fig. 10-2 is shown a simple process heat exchanger. In this heat exchanger, a liquid flows through the exchanger and is heated by condensing steam. The controlled variable for this simple example is the exit temperature of the liquid flowing through the exchange. The manipulated variable is the steam flow to the exchanger and it has its own cascaded feedback control loop. Basically, feedforward control is needed to calculate the desired value of this manipulated steam flow. The significant disturbances to this particular process are assumed to be the inlet liquid temperature and the flow rate of liquid through the exchanger.

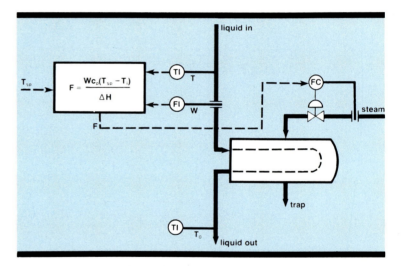

Fig. 10-2. Feedforward Control of a Heat Exchanger

A steady-state energy balance around this exchanger gives the following equation:

$$WC_p (T_0 - T_i) = F\Delta H \qquad (10\text{-}1)$$

where:

W = liquid flow, lb/hr
C_p = liquid heat capacity, Btu/lb -°F
T_i = liquid inlet temperature, °F
T_0 = liquid outlet temperature, °F
F = steam flow, lb/hr
ΔH = heat released by steam, Btu/lb

Equation (10-1) can be solved for F:

$$F = \frac{WC_p(T_0 - T_i)}{\Delta H} \tag{10-2}$$

In this basic equation for F the controlled variable T_0 appears. Replace T_0 with its setpoint or desired value T_{sp} and the result is an equation that gives the steam flow necessary to produce the desired outlet temperature:

$$F = \frac{WC_p(T_{sp} - T_i)}{\Delta H} \tag{10-3}$$

This particular equation has W and the inlet temperature T_i measured and sent to the feedforward controller. Reasonable values for C_p and ΔH should be readily available and are entered by the operator. With this information and the desired value of the outlet temperature T_{sp} it is possible to calculate the steam flow F and feedforward control can be implemented as shown in Fig. 10-2.

This particular example illustrates that the feedforward control equation for a specific loop is uniquely designed. Note the significant difference to feedback control where standard control equations usually are combinations of three specific modes. To be able to design a particular feedforward algorithm for a particular process control application requires significant understanding on the part of the designer. In addition, it means the controller hardware used in feedforward control will be unique in each installation.

Inspection of Fig. 10-2 will show that if there are errors involved in the computation or if there are other disturbances at work that are unforeseen, the control hardware will have no way of sensing the resultant change in outlet temperature of the liquid and, as a result, no effective corrective action will be taken.

10-3. Steady-State or Dynamic Feedforward Control?

The example in the previous section was a specific application of steady-state feedforward control, i.e., when the basic energy balance was made around the heat exchanger, it was a steay-state energy balance. It did not give any quantitative recognition of the fact that there are process dynamics associated with the operation of the exchanger. But in practice, when the inlet conditions change, the resultant changes in the controlled variable will not occur instantaneously. Yet in the specific feedforward control equation, it was implicitly assumed that instantaneous correction was appropriate. When any condition changed in the feedforward control loop, there was an instantaneous (algebraic) calculation of a new steam flow.

For increased accuracy in feedforward control, it is desirable and often necessary to include the effects of process dynamics in the feedforward adjustment of the manipulated variable. This can be done by one of two general, overall strategies. In one case, it is possible to implement steady-state feedforward control and simply *push the output signal* through some type of dynamic compensator before that signal is used to adjust the manipulated variable itself. This type of dynamic compensation is illustrated in Fig. 10-3.

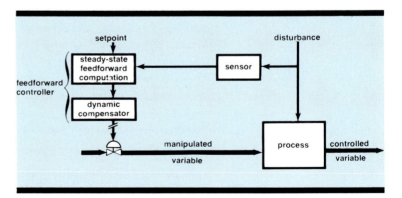

Fig. 10-3. Dynamic Compensation

With dynamic compensation of the type illustrated in Fig. 10-3, it is possible to manually adjust or tune the dynamic compensator so that it introduces proper dynamic corrections to the feedforward control action taken. In typical hardware applications, quite often the dynamic compensator is a simple lag-lead network or a simple second-order lag adjustment. In such cases, the practitioner tunes the ratio of the lag-lead time constants or tunes the ratio of the time constants in the second-order lag network. Such tuning is done after hardware installation.

The second, and more theoretically sophisticated, way to provide dynamic compensation is through the general derivation of the feedforward control equation as a dynamic or unsteady-state relationship. This will be illustrated in the next section.

10-4. General Feedforward Control

It is possible to structure feedforward control in a more general sense than has been done previously. Refer to Fig. 10-4 which shows a process to be controlled in a feedforward manner.

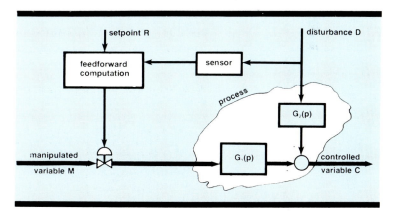

Fig. 10-4. Generalized Feedforward Control Block Diagram*

*The symbol G(p) implies the block diagram symbol for the dynamic relationship between the input to the block and the output from the block. p is the differential operator implying d/dt, i.e., it implies taking the derivative with respect to time. The implication is that the overall relationship may be a differential equation.

In looking at Fig. 10-4, it is seen that the controlled variable may be expressed in terms of the two input forces working on the process, i.e., the disturbance D and the manipulated variable M. This relationship follows:

$$C = G_1(p)M + G_2(p)D \qquad (10\text{-}4)$$

This may be solved for the manipulated variable M:

$$M = \frac{C - G_2(p)D}{G_1(p)} \qquad (10\text{-}5)$$

But in every case of pure feedforward control, you do not actually measure and feed back the controlled variable C; instead the reference or desired value R is taken and substituted for C in the overall control equation. This gives:

$$M = \frac{R - G_2(p)D}{G_1(p)} \qquad (10\text{-}6)$$

This is the general feedforward control equation. D is measured and the setpoint R is provided. We must provide in advance the specific relationships $G_1(p)$ and $G_2(p)$. A continuous calculation is then made of the needed value of M that will maintain R equal to C. The significant factor, however, is that the designer of this control system must have enough understanding of the process so that he can make a specific quantitative determination of the mathematical relationship between D and C, which is $G_2(p)$, and the mathematical relationship between the manipulated variable M and the controlled variable C, which is $G_1(p)$. The major increase in theoretical understanding required to implement feedforward control is clearly seen.

It should be understood that if there are two significant disturbances entering the process, each of them must have its own specific feedforward control computation and, by implication, there will be an additional control equation similar to Eq. (10-6). The difficulties in handling multiple disturbances make feedforward control impractical to implement on a very broad scale.

10-5. Combined Feedforward and Feedback Control

It has been seen that pure feedforward control has some significant disadvantages, i.e., the computations must be sophisticated and take into account many effects; sometimes these involve sophisticated mathematical relationships. When errors are made in modeling or computation, there is no corrective action. In addition, if disturbances other than the specific ones being measured for the feedforward controller enter the process, the automatic control system does nothing and errors build up. As a result of these several problems, pure feedforward control is never encountered by itself; there is always some sort of feedback control added to supplement the feedforward arrangement. Several examples will illustrate how this is done.

Fig. 10-5 shows the simple heat exchanger of Fig. 10-2, but also shown are two ways to couple feedback control with the feedforward arrangement. In Fig. 10-5(a), a feedback controller is used to bias the output of the feedforward control output. In effect, a constant (or *fudge* factor) K_f can be added to our basic computation for steam flow to give:

$$F = WC_p(T_{sp}-T_i) / \Delta H + K_f \qquad (10\text{-}7)$$

In Fig. 10-5(a), the feedback controller will manipulate the value of K_f necessary to maintain $T_0 = T_{sp}$.

Fig. 10-5(b) shows an alternate way to use a feedback control arrangement to supplement the feedforward control of the heat exchanger. In this particular case, the output of the feedback controller is used to adjust the setpoint of the feedforward controller.

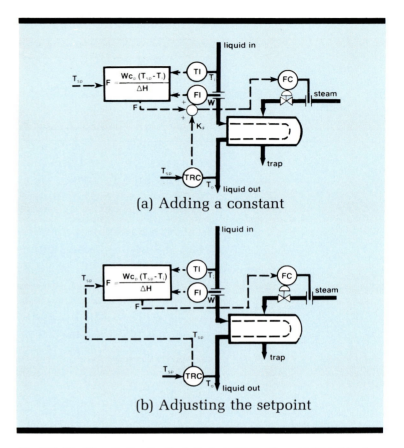

Fig. 10-5. Two Ways To Use Feedback Control with Feedforward Control for A Heat Exchanger

The examples in Fig. 10-5 illustrate feedforward control combined with feedback control, and there also is shown a cascade arrangement to provide the control of the manipulated variable. This cascade aspect of the control scheme is common when feedforward control is used. It grows out of a general acceptance of the fact that if the variable is important enough to merit feedforward control, it is also inherently important enough to merit a cascaded arrangement to insure that the manipulated variable is maintained at its calculated value.

One of the most frequently encountered examples of combined feedforward and feedback control is in the control of water level in a boiler steam drum. This is illustrated in Fig. 10-6, one of the classic *three-element* regulators:

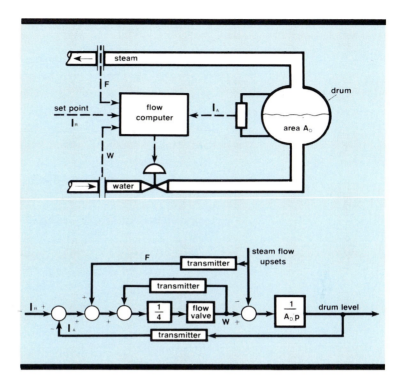

Fig. 10-6. A Three-element Feedwater Regulator

Steam flow from the steam drum is measured and this *load* on the system is sensed and compensated by using feedforward control. Feedback control on the water level in the drum is handled in the conventional manner, and the manipulated flow into the steam drum (make-up water) is controlled through a cascade arrangement. This general arrangement is feedforward plus feedback plus cascade control and, hence, the name three-element feedwater regulator. This is a classic example of this type of control and there are literally thousands of these three-element regulators installed.

In such an arrangement, feedforward control, in effect, is used to provide compensation for significant variations in the major disturbance or the major load on the process. In arrangements such as this (and Fig. 10-6 is a good example), many practitioners refer to the overall control arrangement as *load compensation*.

10-6. The Multivariable Control Problem

Previous units viewed the control system as being 1 x 1, i.e., there was a single controlled variable and a single manipulated variable for the process. But, in practice, this is seldom the case. Quite often, a process is found which involves more than one controlled variable and, therefore, more than one manipulated variable. This is illustrated in Fig. 10-7 for a 2 x 2 process (two controlled variables and two manipulated variables).

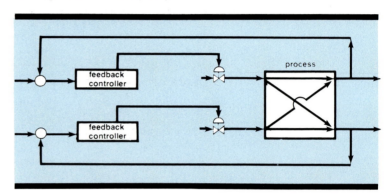

Fig. 10-7. Conventional Multivariable Arrangement

When there are two control loops on one unit, quite often the two control loops will *interact* with one another, i.e., when there is a change in the manipulated variable in loop 1, not only will it produce a change in the controlled variable for loop 1 but also it will produce a change in the controlled variable for loop 2. Conversely, when there is a change in the manipulated variable for loop 2, not only will it produce a change in the controlled variable for loop 2, but also in the controlled variable for loop 1. This type of control interaction produces significant operating problems. These problems of interaction are illustrated in Fig. 10-8.

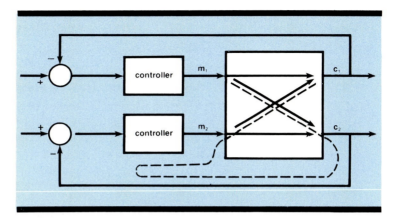

Fig. 10-8. The Effects of Interaction in a 2 x 2 System

In Fig. 10-8, when there is a change in the manipulated variable for loop 1, there will be a change in the controlled variable for loop 1, but, in addition, the change in the manipulated variable for loop 1 will produce a change in the controlled variable for loop 2. This, of course, will be sensed, fed around loop 2 and will, in turn, produce a change in the manipulated variable for loop 2. This not only produces the change in the controlled variable for loop 2, but also produces a further change in the controlled variable for loop 1. In effect, these phenomena start to chase one another in a figure-eight type path. To compensate for this, it is necessary for the operator to desensitize the loops and it becomes very difficult to obtain quality feedback control.

10-7. Implementing Multivariable Control

When feedback loops are interacting with one another, a control system is needed that will *decouple* the loops, i.e., the interaction between the loops needs to be broken. The overall concept of breaking this interaction is referred to as *multivariable control*. The basic approach is to use conventional feedback control supplemented by a *decoupler*. This is illustrated in Fig. 10-9.

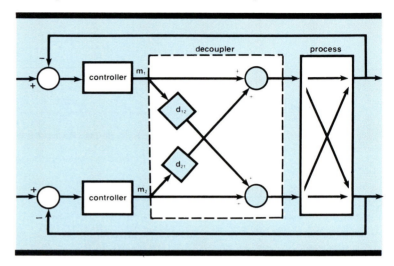

Fig. 10-9. A Decoupler for a 2 x 2 Process

In effect, a decoupler such as shown in Fig. 10-9 is essentially two feedforward control elements. The overall operation is as follows: When a change in the controller output for loop 1 occurs, it not only produces a change in the manipulated variable for loop 2, so that when the interaction from loop 1 to loop 2 takes place, the change in the manipulated variable in loop 2 will exactly compensate for this and, as a result, there will be no change in the controlled variable for loop 2. Conversely, in loop 2 when there is a change in the controller output, it not only will produce a change in the manipulated variable for loop 2, but also it will produce a change in the manipulated variable for loop 1 so that when the interaction from loop 2 to loop 1 takes place, it will be compensated by the change in manipulated variable in loop 1 and there will be no change in the controlled variable in loop 1.

The concept involved in the decoupler shown in Fig. 10-9 is very appealing; the difficulty is in the design of the decoupler itself. The goal is to design the decoupler so that all interactions will be compensated exactly. This is a laudable goal, but often difficult to achieve in practice. For a 2 x 2 process, it is equivalent to the design of two simultaneous feedforward controllers. The problem escalates rapidly, however, and if the system were a 5 x 5 process, the implementation of such multivariable control would require the implementation of 5^2 - 5 or 20 simultaneous feedforward controllers. The increasing difficulty is obvious.

Quite often a decoupler such as shown in Fig. 10-9 is implemented as a steady-state decoupler, i.e., one consisting of algebraic terms only. It is then possible later to add dynamic compensation to the more important elements of the decoupler and to gain further improvement.

You should begin to suspect by now that multivariable control represents a most powerful and most difficult type of automatic control to implement. It requires significant design expertise and capability, and, as a result, it also requires custom hardware capabilites. This conventionally implies the necessity of digital computation capabilities in the hardware. The cost of designing for multivariable control can be quite significant, but the improvement in control also can be significant. There are increasing numbers of commercial applications of multivariable control that have excellent economic justification.

Exercises:

10-1. *Refer to Fig. 7-5. Design a steady-state feedforward control system for this unit.*

10-2. *To Exercise 10-1, add a lag-lead dynamic compensator of the form:*

$$\frac{1 + \tau_1 p}{1 + \tau_2 p}$$

Where the ratio of τ_1/τ_2 is adjustable (tunable).

10-3. *Add cascade control of the manipulated steam flow in Exercise 10-2.*

10-4. *Refer to the results of Exercise 9-3. Install a flow meter on the feed input to the column, maintain reflux flow rate as a ratio of column feed flow, send signal from feed orifice through the ratio controller and then through a dynamic compensator. Now use the composition controller to adjust the ratio.*

10-5. *Given the 2 x 2 control system below:*

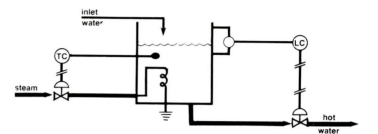

Draw the functional block diagrams for these two loops.

10-6. *Sketch the functional outline of a decoupler for the system of Exercise 10-5.*

Unit 11:
Digital Control

UNIT 11

Digital Control

The development of digital computer capabilities has had a truly revolutionary impact on the hardware available for process control. This in turn has produced major changes in the way in which process control theory is practiced in industry. It is most important that you appreciate these phenomena.

Learning Objectives — **When you have completed this unit, you should:**

A. **Understand the role of digital computers in automatic process control systems.**

B. **Know the meaning of direct digital control and supervisory digital control.**

C. **Be able to explain the concept of distributed control.**

11-1. Digital Capabilities

During the late forties, the electronic digital computer came into existence. Since then it has expanded information processing capabilities in a revolutionary manner. Today in the United States, an incredible digital capability for the processing and handling of information permeates every aspect of life. The data are imprecise, but it has been estimated that if the capabilities of all of the digital computers in the United States were totaled, it would be possible to do several billion computations once per hour for every man, woman, and child living in the United States. This is an amazing capability—one that has come into being because of its practical economic utility. The growth of digital capabilities has had a revolutionary impact on process control and the way in which process control is practiced. (Simply describing the effects as *revolutionary* is probably an understatement.)

The significant question is how have digital computer capabilities been applied (and how should they be applied) to increase the profitability of a process unit. The computer has been used frequently in the past to replace hardware already in existence, i.e., the digital computer became the way to automate (really to reautomate) a function that was already being accomplished with some other type of hardware. But it also has been apparent that the computer can be used to do many things that were not automated before. Digital computers have opened vast, new vistas of usage, and new ways are still being found to exploit their capabilities. They have caused a revolution in the practice of process control theory that is still going on. It also will be many years before the full extent of changes that are occurring today are understood and appreciated.

Digital control is a vast subject in itself, and thus, the coverage in this unit is limited. ISA publishes a separate ILM on digital control which the student will find helpful.

11-2. Digital Control History

Some of the first commercial applications of digital computers in automatic process control installations occurred during the late fifties. The first efforts at the use of digital computers in process control were sometimes crude and quite often unprofitable. The hardware and the concepts involved appear elementary when compared to today's very sophisticated and capable computer installations.

Many of the first uses of digital computers in process control came from the adaptation of conventional computers to sensor-based and real-time oriented systems, i.e., computers that were designed for doing business work such as inventory controls and scientific computations were provided with *front-end* hardware to receive sensor signals. They were used to make computations quickly enough so that they could be used in on-line control to affect the ongoing operation of the process. This concept of real-time operations grew and developed during the early sixties. In time, computers with an architecture specifically designed for real-time capabilities and for process control were developed and marketed. They were faster and architecturally superior to the earlier EDP machines that had been modified with facilities for sensor input. The revolution was continuing and developing a momentum of its own when it was shocked during the late sixties by the introduction of minicomputers; in short order, there were scores of companies offering small, powerful, inexpensive pieces of digital hardware for specific individual control tasks. Many process control applications for digital computers were reduced to manageable project size and a tremendous boost in the use of computers in process control occurred.

The use of minicomputers in process control was just beginning to be understood when a whole new phenomenon was created with the introduction of microprocessors. These inexpensive and powerful digital devices have revolutionized not only digital control but also more conventional analog control. Digital capability becomes practical for even the most mundane and routine task, and the use of digital capability has begun to permeate every aspect of control hardware. Developments show no signs of abating and the revolution continues strong even today; the rate of change is staggering.

It will be helpful to move away from the hardware, however, and look at the functional way in which computers are used in process control.

11-3. Data Logging

Digital computers can be used to log data about the process. This concept is shown in Fig. 11-1 illustrating a conventional process control system at work. A digital computer is used to log data about the process and its operating conditions.

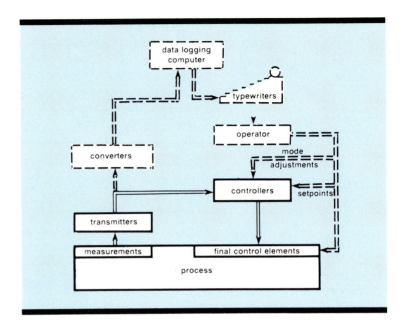

Fig. 11-1. A Data Logger

The use of digital computers for data logging was quite popular during the sixties and many such systems were sold on the basis that a systematic, uniform, continuous logging of data about a process would lead the operators and supervisors to wiser and more significant understanding of the nature of the process and how it should be controlled and managed. However, the distinction between understanding and data soon came into play and a significant disillusionment with data logging developed. This continued until the seventies, but then data logging regained meaningful usage with the availability of cheaper computer hardware and more sophisticated understandings of the role of data logging in plant operations.

There are many situations in which the data associated with the operation of a plant are a significant part of the product of the plant's operation. One obvious example is in pilot plant situations; in addition, many situations exist in which logs on the operating conditions of the plant have significant statutory, regulatory, or legalistic implications. In addition, questions of safety, product reliability, and product specifications are supplemented and assisted by using the digital computer for the routine and systematic collection of data. The use of digital computers for data logging seems to have achieved a regular and meaningful role. It is clearly a part of the broader scene of the use of computers in the overall control and operation of processing units.

11-4. Direct Digital Control (DDC)

The controller in a conventional analog feedback system performs rather routine calculations on the error signal and determines what the value of the manipulated variable should be. The feedback controller is a special-purpose analog computer. As digital computers developed in capability and availability, it became apparent that they could be used to perform the calculations that were being done by analog feedback controllers. In effect, the idea developed that a single digital computer could be time-shared among a number of different feedback control loops; this is illustrated in Fig. 11-2.

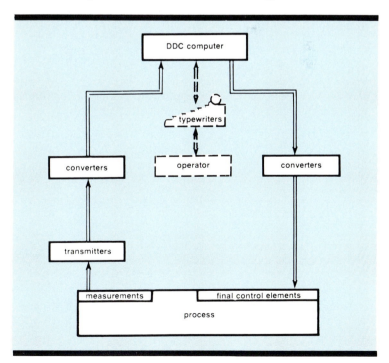

Fig. 11-2. Direct Digital Control (DDC)

The computer has an input subsystem which provides for collection of signals from the various sensors of the controlled variables and other significant operating conditions of the plant. The digital computer calculates—using conventional feedback control equations or algorithms—the needed values of the manipulated variables in the plant, and the computer output becomes signals to the various final control elements.

Basically, in this elementary concept of direct digital control or DDC, the computer is used as a discrete equivalent to the analog hardware which it has replaced. For the conventional three-mode PID controller, the computer equation used to calculate the manipulated variable is a discrete equivalent of the conventional three-mode algorithm. In finite difference form, it is:

$$m_n = K_c e_n + \frac{K_c T}{T_i} \sum_{i=0}^{n} e_i + \frac{T_d K_c}{T} (e_n - e_{n-1}) + M_r$$

$$(11-1)$$

where: m_n = value of manipulated variable at the *nth* sampling instant

e_n = value of the error at the *nth* sampling instant

T = sampling time

M_r = midrange adjustment

The other parameters are all as defined earlier.

This particular form of the direct digital control algorithm is referred to as the *position form* of the control algorithm, and its output is the actual value of the manipulated variable, e.g., a valve position. If this equation were rewritten for the m_{n-1} sampling instant and subtracted from m_n as given above, the result is:

$$\Delta m_n = K_c\,(e_n\text{-}e_{n\text{-}1}) + \frac{K_c T}{T_i}\,e_n + \frac{T_d K_c}{T}\,(e_n\,\text{-}2e_{n\text{-}1} + e_{n\text{-}2})$$

$$(11\text{-}2)$$

$\Delta m_n = (m_n\text{-}m_{n\text{-}1})$ and this is the needed *change* in the manipulated variable. This is referred to as the *velocity algorithm*. The significant difference is that the velocity algorithm does not contain the term M_r; therefore, the computer will provide a smooth transition from manual to automatic control, i.e., we will have *bumpless* transfer.

The first ideas for DDC systems were based on the concept that one computer could be the replacement for many analog controllers, and, therefore, the savings in analog hardware could be used to purchase the digital computer. There are several problems in this being sufficient economic justification, however. Problems result from the fact that DDC systems tend to be more expensive, and programming costs quite often turn out to be large. In addition, some analog backup hardware is required since operating personnel must be able to exercise effective control over the plant in the event of computer failure. In some cases, this analog backup becomes a complete analog control system. Reduction in manpower is a potential justification for DDC systems, but this rarely happens. Most processing units are already automated and operating with a minimum staff, and therefore, few people can be eliminated. As a matter of fact, the presence of the computer frequently entails the presence of higher-caliber personnel and personnel costs often increase rather than decrease.

The sum of much experience in DDC leads to the conclusion that if the computer is used solely to replace earlier automation, it is rarely justified. Saying this differently, the computer needs to be used to automate functions and operations that were not being automatically accomplished before. The truly remarkable capabilities of the computer should be used to solve operating problems and improve the optimization of the unit. If this can be done, i.e., if the capability of the computer can be exploited, then DDC quite often becomes economically attractive. Many of the concepts outlined in the previous two units, e.g., feedforward control, dead time control, multivariable control, etc., can be conveniently implemented using DDC hardware systems. To accomplish many of these with conventional analog hardware would be difficult or impossible.

This broad reasoning leads to the conclusion that the best DDC approach is to implement direct digital control only for those loops in which there is significant improvement in control performance by doing so. The implication is that conventional analog control systems should be left on the loops where conventional feedback control is envisioned. This is a type of "hybrid" approach to process control because it involves the mixing of digital capabilities with conventional analog capabilities.

11-5. Supervisory Digital Control

When reviewing the conventional process control unit, you see that basically a plant operates with large numbers of feedback controllers and occasionally you encounter a particular loop with a specific piece of feedforward control or other sophisticated or advanced control hardware in place. But, even with all of this control hardware in place, closer inspection reveals the fact that the supervisory function of the plant usually is not automated at all. As an example, the individual operators or plant supervisory personnel make the determination of all of the setpoints for the feedback and feedforward controllers. They do this based on past experience and some elementary calculations. Experience has shown that quite often operating personnel undertake this responsibility in a very crude, very conservative, and very slow fashion. The result is that most plants are not operated in their optimal manner.

The basic objective of a process operation is to produce financial return (profit) on investment. The economic return depends upon a number of factors and the operating strategy is not always clearly perceived or broadly understood by all of the decision-making people involved. A plant is a complex, highly interactive entity, and many times the optimum operating strategy can be determined only through sophisticated calculations. Clearly the digital computer should be able to assist in performing such calculations in a real-time, optimal manner. The concept of this is illustrated in Fig. 11-3.

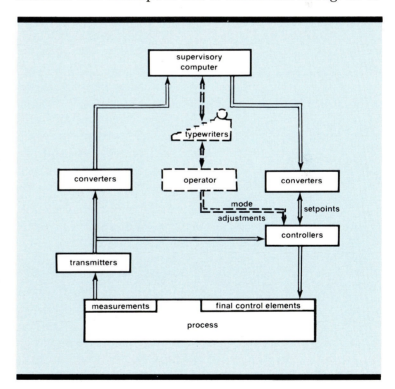

Fig. 11-3. Supervisory Digital Control System

The computer is able to measure significant information about actual operating conditions in the plant. Additional information may be provided to the computer such as:

- the cost of raw materials and utilities;

- the value of products;

- the composition of various raw materials, products, and intermediate streams;

- the current values of variables within the process;

- constraints on the operation of the process;

- specifications on products, etc.

A model of the process can be used to relate these various factors, including a complete picture of plant economics. This model can then be *optimized* to determine the best operating strategy for the unit. The result is a specific indication of the optimum or most desirable values for the setpoints of all of the individual controllers within the plant. These setpoints may be provided simply as information to the operator, i.e., the supervisory control system may be *open loop*. In other cases, the setpoints for the controllers may be set directly in via a *hard-wired* system. This is referred to as *closed-loop* supervisory control.

The installation of supervisory control systems has proven to be an economically attractive concept in process control. It illustrates the fact that this function was not automated before digital computers entered the scene, and it often was being done less well than it can be done by the computer.

The economic justification of supervisory control systems is based on the prospects of the control system producing improvement in process operations. There are, therefore, a number of situations where supervisory digital control systems are prime candidates for consideration. These include:

- plants with large throughput;

- very complex plants;

- plants subject to frequent disturbances, etc.

The major difficulty in the installation of supervisory control systems is that mathematical models of plants are seldom available beforehand, and, therefore, significant engineering and technical investigation are required before the system may be installed and placed in operation. The incentive is clearly present, however, because of the significant benefits that can accrue.

11-6. The Hierarchy Concept

The preceding sections have presented two distinct approaches to the use of digital computers in process control, and any actual digital system can certainly be a hybrid of the two, containing the best aspects of each. It also becomes clear that in any particular processing unit, it is not a question of whether DDC or supervisory control is used; there may very well be situations which use both, plus having some portions of the plant on conventional analog control. In such cases, there is value in functionally viewing the control scheme as being placed in a hierarchial arrangement. This, illustrated in Fig. 11-4, which shows the overall potential.

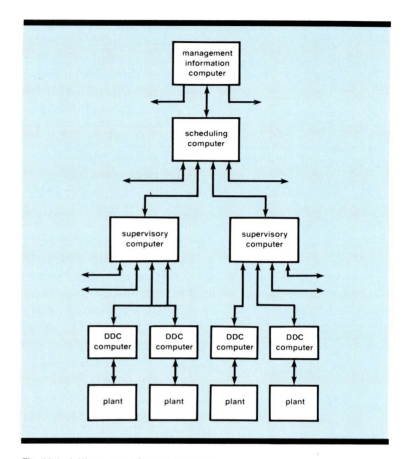

Fig. 11-4. A Hierarchy of Control Systems

This overall concept is perhaps extreme, but it does illustrate the hierarchial framework in which individual applications of digital capabilities might be undertaken. Conceptually, more and more people are beginning to question this sort of *layered* concept of control. The feeling is growing that perhaps the use of computers in control is structured as a bureaucracy. This uneasiness—coupled with the development of cheap microprocessors—has led to the advanced concept of *distributed control* which will be explored in more detail.

11-7. Distributed Control

The hierarchial concept of digital control which was presented in the previous section gained much attention during the late sixties and during the seventies. Toward the end of the seventies, however, the concept of distributed control became a reality. Basically, the concept of distributed control is an architectural view of the structure of a plant's total control system. The concept of distributed control owes its existence to the inexpensive availability of digital microprocessors—itself a phenomena of the late seventies.

Distributed control implies that the actual control and management functions are, in fact, distributed throughout the entire plant—they are not concentrated in a specific geographic location called a control room or in a single central computer. Since the need for control and management is distributed, it is desirable to distribute the hardware and the capabilities of accomplishing this control and management. In order to be able to do this, it is important that there be excellent data communication links throughout the operating unit, that there be digital control hardware and analog control hardware available for use throughout the plant, that all of these capabilities be able to stand the various environmental restrictions placed upon them, and that they be able to work together as a system. In addition, there is the obvious necessity for sophisticated software to coordinate the many communication functions.

Distributed control, as mentioned earlier, is an important architectural view and, as such, merits more extensive treatment. This will be accomplished in the next unit.

Exercises:

11-1. *Why would the development of digital computer technology have a greater impact on process control and management than on other aspects of process equipment?*

11-2. *Discuss the concept of operating data being considered as part of the **product** of a plant, especially a pilot plant.*

11-3. *T is sample time in the DDC algorithm. The sample time for various types of feedback loops has been recommended as follows:*

- *flow loops—once per second*

- *level and pressure loops—every five seconds*

- *temperature loops—every twenty seconds*

Why would the recommendations vary from loop to loop?

11-4. *What is the effect of very large sample times if you **sample and hold** a signal as shown below?*

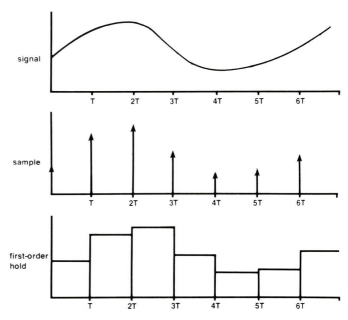

11-5. *Supervisory digital control tends to modify the role of the operator/supervisor. It also tends to bring special attention to the man-machine interface. Define, by sketch if you wish, the nature of the human supervisor's role in a control room.*

11-6. *Supervisory digital control implies optimization and the scope of the optimization may vary from a single item of equipment to an entire plant. What effect does this have on computer hardware needs? What effect on user programming requirements?*

Unit 12:
Automation System Concepts

UNIT 12

Automation System Concepts

The overall automation of an industrial process involves not only designing the control system, but also establishing plant operating procedures and, hence, plant economics. It is important that you gain some insight into and appreciation of the broad philosophical and conceptual framework that forms the basis of the architecture of control systems. This unit describes the functions to be accomplished, the technologies required, and the manner in which the system can be structured.

Learning Objectives
— **When you have completed this unit, you should:**

A. Have developed an appreciation of the various characteristics of increasing levels of process control and process management.

B. Have developed some understanding of the technologies required for effective process control and management.

C. Developed an understanding of basic architectural philosophies for process automation, especially understanding distributed control.

12-1. Process Control and Process Management

This ILM has been devoted principally to specific control techniques and hardware, with little attention given to the overall structure of process automation. In Unit 2, briefly covered is the distinction between process control and process management, and it is now appropriate to expand this further.

When a process is automated, the first general efforts are toward the measurement of process variables and simple hardware automation techniques are used to establish basic control over the operation of the plant by controlling a few specific variables. As process control functions become more elaborate and higher levels of plant automation are undertaken, there begins to be a shift of focus toward automating more and more of the management of the plant.

It is desirable to conceptually separate the process control function from the process management function. It is realized that quite often the distinctions between these two basic areas of automation are blurred and diffuse, but, by appreciating some distinctions between them, you gain a better overall understanding of total plant automation.

Shown in Fig 12-1 is a general outline of the progression of plant automation control systems. The vertical axis in Fig. 12-1 shows increasing levels of control automation while the horizontal axis shows increasing levels of process management. It is clear that the major initial thrusts of automation are toward plant control, but subsequent and advanced control systems begin to focus more and more on automating the management function.

12-2. Specific Characteristics of Process Control Automation

The vertical scale in Fig. 12-1 describes increasing levels of control automation, and it is necessary to understand in more detail the characteristics of the five levels defined. The five general levels of control automation in Fig. 12-1 have specific individual characteristics as shown in Table 12-1(a). As the change is made from manual operation to fully automatic operation, the various characteristics of process control become increasingly automated and more sophisticated.

As a process becomes more fully automated, it is necessary to bring more and more technology into use. In Table 12-1(b) the increasing levels of control automation from Fig. 12-1 are presented to show the functional characteristics of the control technologies required.

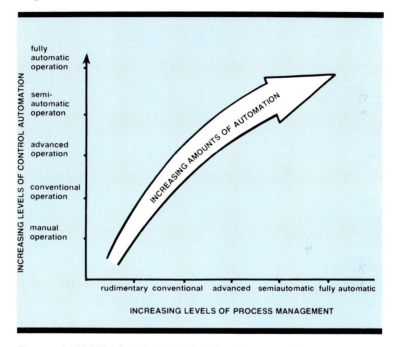

Fig. 12-1. PROGRESSION OF ADVANCED CONTROL SYSTEMS
(Courtesy Foxboro Company)

Table 12-1(a)
CHARACTERISTICS OF INCREASING LEVELS OF CONTROL AUTOMATION (Courtesy Foxboro Company)

CHARACTERISTICS / LEVELS OF CONTROL AUTOMATION	Manipulation Of Process Actuators — Under Normal Operation	Manipulation Of Process Actuators — During Startup, Shutdown Level Chgs.	Manipulation Of Process Actuators — During Emergency Conditions	Control Law Used During Normal Operation Of Plant	Method Of Controller Tuning	Human Monitoring And Intervention	Alarm Condition Analysis	Measurements Used And Variables Computed	Plant Scope
Manual Operation	Human	Human	Human	Written Decision Rule	NA	NA	Human	As Necessary	Independent Unit
Conventional Operation	Automatic		Key Actuators Automatic	On-Off, 3-Mode, Cascade	Human	Automatic Manual, Emergency Trip			
Advanced Operation				Above + Feedforward, Noninteracting	Some Key Loops Tuned Automatically		Some Alarm Analysis Displayed		Integration Of Single Units
Semi-Automatic Operation		Semi-Automatic	All Actuators Automatic	Above + Coordinated	All Key Loops Automatic		Some Automatic Action		Integration Of Multiple Units
Fully Automatic Operation		Fully Automatic			Fully Automatic	Emergency Trip	Fully Automatic		Plant Treated As Single Entity

Table 12-1(b)
FUNCTIONAL CHARACTERISTICS OF CONTROL TECHNOLOGIES REQUIRED (Courtesy Foxboro Company)

FUNCTIONAL CHARACTERISTICS OF CONTROL TECHNOLOGIES REQUIRED / INCREASING LEVELS OF CONTROL AUTOMATION DESIRED	MANUAL OPERATION	CONVENTIONAL OPERATION	ADVANCED OPERATION	SEMI-AUTO. OPERATION	FULLY AUTO. OPERATION
MEASUREMENTS					
Basic Measurements (T, P, L, F)	X	X	X	X	X
Advanced (Composition, Vibration)			X	X	X
Complex (Product Quality)				X	X
Two State Detectors				X	X
CONTROL DECISIONS					
On-Off Controller		X	X	X	X
3-Mode Controller		X	X	X	X
Model-Based Controller			X	X	X
Alarm Diagnostic Routines			X	X	X
Automatic Controller Tuning Routine			X	X	X
Logic Sequencing				X	X
DISPLAY AND REPORTING					
Display and Recording of Variables	X	X	X	X	NA For Normal Operation
Annunciate Alarms		X	X	X	
Alarm Diagnostic Messages			X	X	
Status Displays				X	
CONTROL MODELS					
Simple Heat & Mat. Balance Models			X	X	X
Dynamic Process Models				X	X
Detailed Safety Models				X	X
VARIABLE MANIPULATION					
Actuators (Basic)	X	X	X	X	X
Actuators (Increased Accuracy)				X	X
Actuators (On-Off)				X	X
Actuators (Increased Reliability)					X

Tables 12-1(a) and 12-1(b) show the designer what is possible in process control automation and the functional characteristics of the associated technologies. It is clear that some industries are more sophisticated in control automation than others and, of course, within specific industries there are wide variations from plant to plant and from company to company. It is clear, however, that the strong trend is toward increasing levels of control automation and the increased application of the technologies associated therewith.

12-3. Specific Characteristics of Process Management

The distinction has been made between process control and process management, and the typical evolution of increased levels of total automation was illustrated in Fig. 12-1. The previous section illustrated some of the specific details of increased levels of process control, and it is now appropriate to undertake a similar presentation with respect to process management. Shown in Table 12-2(a) are the various levels of process management (as defined in Fig. 12-1), and for each of these the associated characteristics are shown.

Just as in the case of process control, the levels of process management have associated with them various technology requirements and these are illustrated in Table 12-2(b). The trend and the analogy to the situation in process control are clear.

With this overall model for process automation, the individual practitioner can better see and envision the automation of his individual project.

12-4. Overall Control Configuration

It is clear from the previous sections that process automation involves large numbers of individual control and/or management functions.

These functions are, of course, distributed throughout the entire processing unit and the *automation* of a plant involves not only the execution of each of these

Table 12-2(a)

CHARACTERISTICS OF INCREASING LEVELS OF PROCESS MANAGEMENT (Courtesy Foxboro Company)

CHARACTER-ISTICS	DATA GATHERING AND RETENTION			USE OF MANAGEMENT AND PROCESS INFORMATION			
LEVELS OF PROCESS MGT. NEEDED OR DESIRED	Data Gathering And Conversion To Machine Readable Form	Data Type Entered Into Process Management System	Data Retention	Display Of Performance Variables	Basis For Decisions	Degree Of Learning	Plant Scope
Rudimentary Process Mgt.	Manual Gathering	None	Logs, Records	None	Human Intuition	Human	Independent Single Unit
Conventional Process Mgt.	Manual Gathering & Conversion	Some Process	Some System Retention	Some Necessary	Human Judgment		
Advanced Process Mgt.	Some Automatic Gathering & Conversion	Most Process Some Business	Most System	Most Necessary	Some Automatic Decisions	Some Models Updated Routinely	Some Integration Of Single Units
Semiautomatic Process Mgt.	Most Automatic	All Process Most Business		All Necessary	Most Automatic	Semi-Automatic Updating	Integration Of Multiple Units
Fully Automatic Process Mgt.	Fully Automatic	All Data	All System	No Variables Displayed	Fully Automatic	Fully Automatic	Plant Treated As Single Entity

Table 12-2(b)

FUNCTIONAL CHARACTERISTICS OF PROCESS MANAGEMENT TECHNOLOGIES REQUIRED (Courtesy Foxboro Company)

FUNCTIONAL CHARACTERISTICS OF PROCESS MGT. TECHNOLOGIES REQUIRED / INCREASING LEVELS OF PROCESS MANAGEMENT DESIRED	RUDIMEN-TARY	CONVEN-TIONAL	ADVANCED	SEMI-AUTOMATIC	FULLY AUTOMATIC
INFORMATION GATHERING AND CONVERSION					
Process Data Gathering	X	X	X	X	X
Process Data Entry		X	X	X	X
Business Data Gathering And Entry			X	X	X
DECISION ANALYSIS TECHNIQUES					
Material and Energy Balance		X	X	X	X
Computation Of Performance Variables		X	X	X	X
Exception Identification		X	X	X	X
Inventory Analysis			X	X	X
Scheduling			X	X	X
Optimization Algorithms			X	X	X
Forecast Systems				X	X
Process Diagnostics				X	X
DISPLAY AND REPORTING					
Data Logging	X	X	X	X	
Trend Displays		X	X	X	NA
Exception Reporting		X	X	X	For
Receipts, Orders, Shipments Display			X	X	Normal
Pattern Displays			X	X	Operation
User-Directed Hierarchical Display			X	X	
DECISION MODELS					
Safety and Environmental Model		X	X	X	X
Material Balance Model		X	X	X	X
Energy Balance Model		X	X	X	X
Inventory Model (e.g. EOQ)			X	X	X
Scheduling Model			X	X	X
Optimization Model (e.g. LP, NLP)			X	X	X
Distribution & Transportation Model			X	X	X
Forecast Model (e.g. Econometric)				X	X
Diagnostic Model				X	X
PROCESS MANAGEMENT DIRECTIVES					
Operating Guides		X	X	X	X
Plant Operating Conditions			X	X	X
Raw Material Selection			X	X	X
Shipping & Distribution Instructions			X	X	X
Utilities Management			X	X	X
Maintenance Schedules			X	X	X
Raw Material Ordering				X	X

individual steps, but also (and perhaps more importantly) the total collection of these individual pieces of automation into a functional and viable *system*.

There are many different philosophies about the most effective and desirable way to connect all of the functional parts of process automation together into a viable system. In Unit 11 (Digital Control), the concept of hierarchial systems was presented in an elementary way. This type of system is intuitively appealing because of its obvious similarity to the bureaucratic organizational structures so often encountered. It is, however, not the only approach to the *connection* problem, and in many respects, it is not the best solution.

In order to give some insight into the various control configurations that are available, refer to Fig. 12-2 which shows five different organizations for bringing together all of the various control components in a plant automation system. Fig. 12-2(a) shows the hierarchial system that was illustrated in Unit 11. It is also clear that there can be a general network such as shown in Fig. 12-2(b) and, of course, when compared to a hierarchy or other similar specialized organization, such a general network does allow for more redundant and reliable communications among the various components.

The data highway or bus system shown in Fig. 12-2(c) is a very popular concept for many existing process control distributed systems, and in one sense this is a system that is fairly highly centralized (about the highway).

Little comment will be made with respect to stars or to ring systems because these are not in significant use at present.

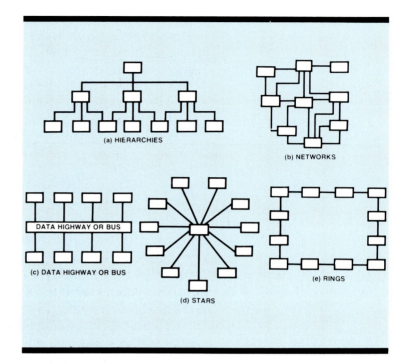

Fig. 12-2. Distributed Organization

After viewing the various distribution-connection concepts available for process automation, it is perhaps desirable to reflect for a moment on why systems are connected at all; there is no physical mandate that this be the case. As a matter of fact, most early process control systems—and the vast majority of those in operation today—are built around the idea of each individual loop operating as a separate, stand-alone entity with no interconnection whatsoever. Even when processes have a central control room of some sort, the loops often are independent, and when loops are connected into a central item(s) of hardware (such as a data gathering computer), they retain their individual operating autonomy. Such an aggregation forms a single, monolithic system, and has little reliability and flexibility.

12-5. Specific Details of Control Configuration

As just stated, for many years all process automation systems had a very simple monolithic form or configuration. Analog signals from individual measurement sensors were transmitted to a central control room and control signals to the actuators were returned in analog form. This basic configuration persisted effectively for three decades—from the time of the first centralized control rooms until the late seventies. Most of the early systems were pneumatic in nature, and, as electronic control systems came on the scene, they tended to duplicate the architectural configuration of pneumatic systems.

The basic concept of this type of centralized control configuration is shown in Fig. 12-3. In such installations, hundreds—and even thousands—of control loops are directly tubed or wired into the control room. When digital control first came into use, much of the discussion was about direct digital control, such as presented in Unit 11. The basic architectural configuration remained monolithic and totally centralized. All information processing and all computation—whether done by analog techniques or digital techniques—were done within the control room or in stand-alone, field-mounted controllers.

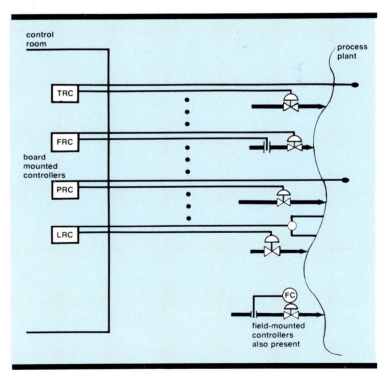

Fig. 12-3. Centralized, Single-Loop Architecture

In one sense, this type of configuration is similar to the *star* type configuration of Fig. 12-2. However, the star configuration of Fig. 12-2 can have elaborate communications in and among the various loops, but this is not the case in the traditional centralized control room arrangement.

It is clear that architectural design has moved away from the simple, individual loop arrangements for process control and has moved toward interconnected distributed systems with elaborate interconnecting communications. The idea of a distributed system is illustrated in Fig. 12-4. In distributed control, the individual feedback controllers for each process loop are removed from the control room location and placed closer to the field sensors and/or actuators. When this is done, of course, there is the need for a significant communication link; e.g., a digital bus or data highway that connects the individual controllers with operators, computers, consoles, and displays. The control loops themselves become physically shorter and therefore less vulnerable to noise or damage. The communication link might be lost, but basically this represents a loss of operator intelligence because the individual field controllers continue to operate locally. Systems of this type can be implemented with either analog or digital controllers and even with pneumatic controllers; but, to date, the mostly digital hardware approach prevails and will continue to do so in the future.

In this approach the operator in the central control room has access to all controller data such as setpoints, process variable measurements, controller output signal levels, etc. Sophisticated displays are also available and the supervisory or management function—with all of the potential for advanced management information, optimization, and supervision—are easily implemented within the control room itself.

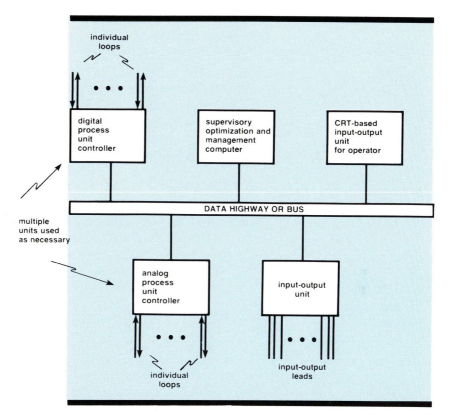

Fig. 12-4. Distributed Architecture

12-6. Summary

The structure of process automation takes on many different forms and architectural details, but the conceptual structure can be visualized as a series of levels of process control and process management. The specific hardware arrangements for the implementation of these individual automation functions vary dramatically from vendor to vendor, and the detailed presentation of any particular hardware configuration is beyond the scope of this particular ILM. It should be clear to the student, however, that process automation is rapidly moving in the direction of broadly distributed, highly interactive, functionally communicative systems. Clearly, this trend is going to continue and accelerate.

As process economic pressures tighten and as constraints on plant operation tighten, the expectations (requirements?) of the process automation system increase. The *window* of appropriate plant operations grows smaller, and better control and automation become a necessity.

Exercises:

12-1. *Review an existing process operating unit and inspect its control system. Now try to place the level of process control and the level of process management that is in place and in effect. Do this by reference to Table 12-1 and Table 12-2.*

12-2. *For the process operating unit of Exercise 12-1, determine what would be required (in terms of technologies) to increase the level of process automation on the unit.*

12-3. *Select two vendors' process automation systems and analyze the hardware's architectural configuration.*

12-4. *Refer to Fig. 12-4. Why would you have both analog and digital process unit controllers?*

12-5. *What effects do increasing energy costs and/or increasing raw material costs have on the level of process automation needed on a unit?*

12-6. *With increasing legal and regulatory constraints on effluent streams, product liability, and product specifications, is there an effect on the level of process automation needed?*

IN CLOSING

You made it through! You have been introduced; you are not yet an expert. This ILM has covered the Fundamentals of Process Control Theory and, hopefully, you now have a more complete overview of the subject.

Appendix A:
Suggested Readings
and Study Materials

APPENDIX A

Suggested Readings and Study Materials

Independent Learning Modules:

One of your best sources of material for further reading and study of process control and instrumentation are the ILMs published by ISA. They are custom designed and created for this exact purpose. Place a *standing purchase order* to receive new ILMs as they are published.

Handbooks and Manuals (selected titles):

Considine, Douglas M. (ed.), *Handbook of Applied Instrumentation*, (McGraw-Hill Book Company, 1964).

Hutchinson, J.W., *ISA Handbook of Control Valves*, 2nd ed. (Instrument Society of America, 1976).

Kallen, Howard P., *Handbook of Instrumentation and Controls*, (McGraw-Hill, 1961).

Liptak, Bela G., *Instrument Engineer's Handbook: Vol. I–Process Measurement Control*, (Chilton Book Co., 1970).

Skrokov, M.R., *Mini- and Microcomputer Control in Industrial Processes: Handbook of Systems and Application Strategies*, (Van Nostrand Reinhold, 1980).

Process Control and Optimization Handbook for the Hydroprocessing Industries, (Gulf Publishing Co., 1980).

Textbooks (selected titles):

Andrew, W.G. and H.B. Williams, *Applied Instrumentation in the Process Industries*, 2nd ed., (Gulf Publishing Co., 1980).

Buckley, P.S., *Techniques of Process Control*, (Wiley, 1964).

Coughanowr and Koppel, *Process Systems Analysis and Control*, (McGraw-Hill, 1965).

Eckman, D.P., *Automatic Process Control*, (Wiley, 1958).

Harriott, Peter, *Process Control*, (McGraw-Hill, 1964).

Hougen, J.O., *Measurements and Control Applications*, 2nd ed., (Instrument Society of America, 1979).

Johnson, E.F., *Automatic Process Control*, (McGraw-Hill, 1967).

Luyben, W.L., *Process Modeling, Simulation, and Control*, (McGraw-Hill, 1973).

Murrill, P.W., *Automatic Control of Processes*, (International Textbook Co., 1967).

Shinskey, F.G., *Process Control Systems*, (McGraw-Hill, 1979).

Smith, C.L., *Digital Computer Process Control*, (Intext Educational Publishers, 1972).

Weber, T.W., *An Introduction to Process Dynamics and Control*, (Wiley, 1973).

Weyrick, R.C., *Fundamentals of Automatic Control*, (McGraw-Hill, 1975).

Technical Magazines and Journals (selected titles):

- *Control Engineering*, published by Dun-Donnelly Pub. Corp., New York, NY
- *ISA Transactions*, published by the Instrument Society of America.
- *Instrumentation Technology*, published by the Instrument Society of America.
- *Instruments and Control Systems*, published by Chilton, Philadelphia, PA

ISA Publications

The Instrument Society of America is a technical, application-oriented society. Its primary goal is providing educational materials and services to its members. If you are seriously interested in the study of process control, you will find membership in the ISA invaluable.

Appendix B:
Glossary of Standard Process Instrumentation Terminology

APPENDIX B

Glossary of Standard Process Instrumentation Terminology

NOTE: The definitions presented in this appendix are all taken from the Instrument Society of America Standard on Process Instrumentation Terminology (ISA-S51.1/1976). This ISA Standard contains many additional terms and there are many notes and amplifying figures that relate to the terminology. The serious student of process control should have a copy of this ISA Standard available for routine reference.

accuracy—Degree of **conformity** of an indicated value to a recognized accepted standard value, or ideal value.

accuracy, measured—The maximum positive and negative **deviation** observed in testing a **device** under specified conditions and by a specified procedure.

accuracy rating—A number or quantity that defines a limit that **errors** will not exceed when a **device** is used under specified **operating conditions.**

actuating error signal—see **signal, actuating error.**

adaptive control—see **control, adaptive.**

adjustment span—Means provided in an instrument to change the slope of the input-output curve. See **span shift.**

adjustment, zero—Means provided in an instrument to produce a parallel shift of the input-output curve. See **zero shift.**

amplifier—A **device** that enables an **input** signal to control power from a source independent of the **signal** and thus be capable of delivering an output that bears some relationship to, and is generally greater than, the **input signal.**

analog signal—see **signal, analog.**

attenuation—1) A decrease in **signal** magnitude between two points, or between two frequencies. 2) The reciprocal of **gain.**

automatic control system—see **control system, automatic.**

automatic/manual station—A **device** which enables an operator to select an automatic **signal** or a manual **signal** as the input to a controlling element. The automatic **signal** is normally the output of a **controller,** while the manual **signal** is the output of a manually operated **device.**

calibrate—To ascertain outputs of a **device** corresponding to a series of values of the quantity which the **device** is to measure, receive, or transmit. Data so obtained are used to:

1. determine the locations at which scale graduations are to be placed;
2. adjust the output, to bring it to the desired value, within a specified tolerance;
3. ascertain the **error** by comparing the **device** output reading against a standard.

cascade control—see **control, cascade.**

characteristic curve—A graph (curve) which shows the ideal values at **steady-state,** or an output variable of a system as a function of an input variable, the other input variables being maintained at specified constant values.

Note: When the other input variables are treated as **parameters,** a set of characteristic curves is obtained.

closed-loop—see **loop, closed.**

closed-loop gain—see **gain, closed-loop.**

compensation—Provision of a special construction, a supplemental **device,** circuit, or special materials to counteract sources of **error** due to variations in specified **operating conditions.**

compensator—A **device** which converts a signal into some function which, either alone or in combination with other **signals,** directs the **final controlling element** to reduce **deviations** in the directly controlled variable.

conformity—Of a curve, the closeness to which it approximates a specified curve (e.g., logarithmic, parabolic, cubic, etc.)

control action—Of a **controller** or a controlling system, the nature of the change of the output affected by the input.

Note: The output may be a **signal** or a value of a **manipulated variable.** The input may be the control loop **feedback** signal when the setpoint is constant, an **actuating error signal,** or the output of another **controller.**

control action, derivative (rate) (D)—**Control action** in which the output is proportional to the rate of change of the input.

control action, floating—**Control action** in which the rate of change of the output variable is a predetermined function of the input variable.

Note: The rate of change may have one absolute value, several absolute values, or any value between two predetermined values.

control action, integral (reset) (I)—**Control action** in which the output is proportional to the time integral of the input; i.e., the rate of change of output is proportional to the input.

control action, proportional (P)—**Control action** in which there is a continuous linear relation between the output and the input.

Note: This condition applies when both the output and input are within their normal operation ranges and when operation is at a frequency below a limiting value.

control action, proportional plus derivative (rate) (PD)—**Control action** in which the output is proportional to a linear combination of the input and the time rate-of-change of input.

control action, proportional plus integral (reset) (PI)—**Control action** in which the output is proportional to a linear combination of the input and the time integral of the input.

control action, proportional plus integral (reset) plus derivative (rate) (PID)—**Control action** in which the output is proportional to a linear combination of the input and the time integral of the input and the time rate-of-change of input.

control adaptive—Control in which automatic means are used to change the type or influence (or both) of control **parameters** in such a way as to improve the performance of the **control system.**

control, cascade—Control in which the output of one **controller** is the **setpoint** for another **controller.**

control center—An equipment structure, or group of structures, from which a **process** is measured, controlled, and/or monitored.

control, differential gap—Control in which the output of a **controller** remains at a maximum or minimum value until the controlled variable crosses a band or gap, causing the output to reverse. The controlled variable must then cross the gap in the opposite direction before the output is restored to its original condition.

control, direct digital—Control performed by a digital **device** which establishes the **signal** to the **final controlling element.** Often called DDC.

control, feedback—Control in which a **measured variable** is compared to its **desired value** to produce an **actuating error signal** which is acted upon in such a way as to reduce the magnitude of the **error.**

control, feedforward—Control in which information concerning one or more conditions that can disturb the controlled variable is converted, outside of any feedback loop, into corrective action to minimize **deviations** of the controlled variable.

control, high limiting—Control in which the output **signal** is prevented from exceeding a predetermined high limiting value.

control, low limiting—Control in which output **signal** is prevented from decreasing beyond a predetermined low-limiting value.

control mode—A specific type of **control action** such as **proportional, integral,** or **derivative.**

control, optimizing—Control that automatically seeks and maintains the most advantageous value of a specified variable rather than maintaining it at one set value.

control, shared-time—Control in which one **controller** divides its computation or control time among several control loops rather than by acting on all loops simultaneously.

control supervisory—Control in which the control loops operate independently subject to intermittent corrective action; e.g., **setpoint** changes from an external source.

control system—A system in which deliberate guidance or manipulation is used to achieve a prescribed value of a variable.

control system, automatic—A **control system** which operates without human intervention.

control system, multielement (multivariable)—A **control system** utilizing **input signals** derived from two or more **process** variables for the purpose of jointly affecting the action of the **control system.**

control system, noninteracting—A **multielement control system** designed to avoid disturbances to other controlled variables due to the **process** input adjustments which are made for the purpose of controlling a particular **process** variable.

control, time proportioning—**Control** in which the output **signal** consists of periodic pulses whose duration is varied to relate, in some prescribed manner, the time average of the output to the **actuating error signal.**

control valve—A **final controlling element,** through which a fluid passes, which adjusts the size of flow passage as directed by a **signal** from a **controller** to modify the rate of flow of the fluid.

control, velocity limiting—Control in which the rate of change of a specified variable is prevented from exceeding a predetermined limit.

controlled variable—See **variable, directly controlled.**

controller—A **device** which operates automatically to regulate a controlled variable.

controller, derivative (D)—A **controller** which produces **derivative control action** only.

controller, direct acting—A **controller** in which the value of the **output signal** increases as the value of the input **(measured variable)** increases. See **controller, reverse acting.**

controller, floating—A **controller** in which the rate of change of the output is a continuous (or at least a piecewise continuous) function of the **actuating error signal.**

controller, integral (reset) (I)—A controller which produces **integral control action** only.

controller, multiposition—A **controller** having two or more discrete values of output.

controller, on-off—A **two-position controller** of which one of the two discrete values is zero.

controller, program—A **controller** which automatically holds or changes **setpoint** to follow a prescribed program for a **process.**

controller, proportional (P)—A **controller** which produces **proportional control action** only.

controller, proportional plus derivative (rate) (PD)—A **controller** which produces **proportional plus derivative (rate) control action.**

controller, proportional plus integral (reset) (PI)—A **controller** which produces **proportional plus integral (reset) control action.**

controller, proportional plus integral (reset) plus derivative (rate) (PID)—A **controller** which produces **proportional plus integral (reset) plus derivative (rate) control action.**

controller, ratio—A **controller** which maintains a predetermined ratio between two variables.

controller, reverse acting—A **controller** in which the value of the **output signal** decreases as the value of the input **(measured variable)** increases. See **controller direct acting.**

controller, sampling—A **controller** using intermittenly observed values of a **signal** such as the **setpoint signal** and the **actuating error signal,** or the **signal** representing the controlled variable to effect **control action.**

controller, self-operated (regulator)—A **controller** in which all the energy to operate the **final controlling element** is derived from the **controlled system.**

controller, time schedule—A **controller** in which the **setpoint** or the **reference-input signal** automatically adheres to a predetermined time schedule.

controller, two-position—A **multiposition controller** having two discrete values of output.

correction time—See **time, settling.**

damping—(1) (noun) The progressive reduction or suppression of oscillation in a **device** or system. (2) (adj) Pertaining to or productive of damping.

dead band—The **range** through which an input can be varied without initiating observable response.

dead time—See **time, dead.**

dead zone—See **zone, dead.**

delay—The interval of time between a changing **signal** and its repetition for some specified duration at a downstream point of the **signal** path; the value θ in the transform factor exp $(-\theta s)$. See **time, dead.**

derivative action time—See **time, derivative action.**

derivative control—See **control action, derivative** (D).

desired value—See **value, desired.**

detector—See **transducer.**

deviation—Any departure from a **desired value** or expected value or pattern.

deviation, steady-state—The system deviation after **transients** have expired.

deviation, system—The instantaneous value of the **directly controlled variable** minus the **setpoint.**

deviation, transient—The instantaneous value of the **directly controlled variable** minus its **steady-state** value.

device—An apparatus for performing a prescribed function.

differential gap control—See **control, differential gap.**

digital signal—See **signal, digital.**

direct acting controller—See **controller, direct acting.**

direct digital control—See **control, direct digital.**

directly controlled system—See **system, directly controlled.**

directly controlled variable—See **variable, directly controlled.**

distance/velocity lag—A delay attributable to the transport of material or to the finite rate of propagation of a **signal.**

disturbance—An undesired change that takes place in a **process** which tends to affect adversely the value of a controlled variable.

drift—An undesired change in the output-input relationship over a period of time.

drift, point—The change in output over a specified period of time for a constant input under specified reference **operating conditions.**

droop—See **offset.**

dynamic gain—See **gain, dynamic.**

dynamic response—See **response, dynamic.**

element—A component of a **device** or system.

element, final controlling—The **forward controlling element** which directly changes the value of the **manipulated variable.**

element primary—The system **element** that quantitatively converts the **measured variable** energy into a form suitable for measurement.

element, reference-input—The portion of the **controlling system** which changes the reference-input **signal** in response to the **setpoint.**

element sensing—The **element** directly responsive to the value of the **measured variable.**

elements, feedback—Those **elements** in the **controlling system** which act to change the **feedback signal** in response to the **directly controlled variable.**

elements, forward controlling—Those **elements** in the controlling system which act to change a variable in response to the actuating **error signal.**

error—The algebraic difference between the indication and the **ideal value** of the **measured signal.** It is the quantity which algebraically subtracted from the indication gives the **ideal value.**

error signal—See **signal, error.**

error, systematic—An **error** which, in the course of a number of measurements made under the same conditions of the same value of a given quantity, either remains constant in absolute value and sign or varies according to a definite law when the conditions change.

excitation—The external supply applied to a **device** for its proper operation.

feedback control—See **control, feedback.**

feedback elements—See **elements, feedback.**

feedback loop—See **loop, closed (feedback loop).**

feedback signal—See **signal, feedback.**

feedforward control—See **control, feedforward.**

final controlling element—See **element, final controlling.**

floating control action—See **control action, floating.**

floating controller—See **controller, floating.**

flowmeter—A **device** which measures the rate of flow or quantity of a moving fluid in an open or closed conduit. It usually consists of both a primary and a secondary device.

frequency response characteristic—The frequency-dependent relation, in both amplitude and phase, between **steady-state** sinusoidal inputs and the resulting fundamental sinusoidal outputs.

gain, closed-loop—The **gain** of a **closed-loop** system, expressed as the ratio of the output change to the input change at a specified frequency.

gain, dynamic—The magnitude ratio of the **steady-state** amplitude of the **output signal** from an **element** or **system** to the amplitude of the **input signal** to that **element** or **system** for a sinusoidal **signal.**

gain, loop—The ratio of the change in the **return signal** to the change in its corresponding **error signal** at a specified frequency.

gain, open-loop—See **gain, loop.**

gain, proportional—The ratio of the change in output due to **proportional control action** to the change in input.

gain, static (zero-frequency gain)—The **gain** of an **element,** or loop **gain** of a system, the value approached as a limit as frequency approaches zero.

hardware—Physical equipment directly involved in performing industrial **process** measuring and controlling functions.

hunting—An undesirable oscillation of appreciable magnitude, prolonged after external stimuli disappear.

hystereis—That property of an **element** evidenced by the dependence of the value of the output, for a given excursion of the input, upon the history of prior excursions and the direction of the current traverse.

I controller—See **controller, integral (reset) (I).**

ideal value—See **value, ideal.**

indicating instrument—See **instrument, indicating.**

inherent regulation—See **self-regulation.**

input—See **signal, input.**

instrument, computing—A **device** in which the output is related to the input or inputs by a mathematical function such as addition, averaging, division, integration, lead/lag, signal limiting, squaring, square root extraction, subtraction, etc.

instrument, indicating—A measuring instrument in which only the present value of the measured variable is visually indicated.

instrument, measuring—A **device** for ascertaining the magnitude of a quantity or condition presented to it.

instrument, recording—A measuring instrument in which the values of the measured variable are recorded.

instrumentation—A collection of instruments or their application for the purpose of observation, measurement, or control.

integral action limiter—A **device** which limits the value of the **output signal** due to **integral control action,** to a predetermined value.

integral action rate (reset rate)—(1) Of **proportional plus integral** or **proportional plus integral plus derivative control action** devices; for a step input, the ratio of the initial rate of change of output due to **integral control action** to the change in **steady-state** output due to **proportional control action.**

Note: Integral action rate is often expressed as the number of repeats per minute because it is equal to the number of times per minute that the proportional response to a step input is repeated by the initial integral response.

(2.) Of **integral control action devices**; for a step input, the ratio of the initial rate of change of output to the input change.

integral control action—See **control action, integral.**

integral controller—See **controller, integral (reset).**

interference, common mode—A form of interference which appears between measuring circuit terminals and ground.

interference, differential mode—See **interference, normal mode.**

interference, normal mode—A form of interference which appears between measuring circuit terminals.

intrinsically safe equipment and wiring—Equipment and wiring which are incapable of releasing sufficient electrical or thermal energy under normal or abnormal conditions to cause ignition of a specific hazardous atmospheric mixture in its most easily ignited concentration.

linear system—See **system, linear.**

linearity—The closeness to which a curve approximates a straight line.

load regulation—The change in output (usually speed or voltage) from no-load to full-load (or other specified load limits). See **offset.**

loop, closed (feedback loop)—A **signal** path which includes a forward path, a **feedback** path, and a **summing point,** and forms a closed circuit.

loop, feedback—See **loop, closed (feedback loop).**

loop gain—See **gain, loop.**

loop, open—A **signal** path without **feedback.**

loop transfer function—Of a **closed-loop,** the **transfer function** obtained by taking the ratio of the **Laplace transform** of the **return signal** to the **Laplace transform** of its corresponding **error signal.**

manipulated variable—See **variable, manipulated.**

modulation—The process, or result of the process, whereby some characteristic of one wave is varied in accordance with some characteristic of another wave.

module—An assembly of interconnected components which constitutes an identifiable **device,** instrument, or piece of equipment. A module can be disconnected, removed as a unit, and replaced with a spare. It has definable performance characteristics which permit it to be tested as a unit.

multielement control system—See **control system, multielement (multivariable).**

multiposition controller—See **controller, multiposition.**

multivariable control system—See **control system, multielement (multivariable)**.

noise—An unwanted component of a **signal** or variable.

noninteracting control system—See **control system, noninteracting**.

normal mode rejection—The ability of a circuit to discriminate against a **normal mode voltage**.

normal mode voltage—See **voltage, normal mode**.

offset—The **steady-state deviation** when the **setpoint** is fixed. See also **deviation, steady-state**.

on-off controller—See **controller, on-off**.

operating conditions—Conditions to which a **device** is subjected, not including the variable measured by the **device**.

optimizing control—See **control, optimizing**.

output signal—See **signal, output**.

overdamped—See **damping**.

overshoot—See **transient overshoot**.

parameter—A quantity or property treated as a constant but which may sometimes vary or be adjusted.

P controller—See **controller, proportional**.

PD controller—See **controller, proportional plus derivative**.

PI controller—See **controller, proportional plus integral**.

PID controller—See **controller, proportional plus integral plus derivative**.

position—Of a **multiposition controller**, a discrete value of the **output signal**.

primary element—See **element, primary**.

process—Physical or chemical change of matter or conversion of energy; e.g., change in pressure, temperature, speed, electrical potential, etc.

process control—The regulation or manipulation of variables influencing the conduct of a **process** in such a way as to obtain a product of desired quality and quantity in an efficient manner.

process measurement—The acquisition of information that establishes the magnitude of **process** quantities.

proportional band—The change in input required to produce a full **range** change in output due to **proportional control action**.

proportional gain—See **gain, proportional**.

range—The region between the limits within which a quantity is measured, received, or transmitted, expressed by stating the lower and upper range values.

rate—See **control action, derivative.**

ratio controller—See **controller, ratio.**

reference-input element—See **element, reference-input.**

regulator—See **controller, self-operated (regulator).**

reliability—The probability that a **device** will perform its objective adequately, for the period of time specified under the **operating conditions** specified.

repeatability—The closeness of agreement among a number of consecutive measurements of the output for the same value of the input under the same **operating conditions,** approaching from the same direction, for full **range** traverses.

reproducibility—The closeness of agreement among repeated measurements of the output for the same value of input made under the same **operating conditions** over a period time, approaching from both directions.

reset control action—See **control action, integral (reset).**

reset rate—See **integral action rate.**

resolution—The least interval between two adjacent discrete details which can be distinguished one from the other.

resonance—Of a system or **element,** a condition evidenced by large oscillatory amplitude, which results when a small amplitude of periodic input has a frequency approaching one of the natural frequencies of the driven system.

response, dynamic—The behavior of the output of a **device** as a function of the input, both with respect to time.

response, ramp—The total (transient plus **steady-state**) **time response** resulting from a sudden increase in the rate of change from zero to some finite value of the input stimulus.

response, step—The total (transient plus **steady-state**) **time response** resulting from a sudden change from one constant level of input to another.

response, time—An output expressed as a function of time, resulting from the application of a specified input under specified **operating conditions.**

reverse acting controller—See **controller, reverse acting.**

rise time—See **time, rise.**

sampling controller—See **controller, sampling.**

sampling period—The time interval between observations in a periodic sampling **control system.**

scale factor—The factor by which the number of scale divisions indicated or recorded by an instrument should be multiplied to compute the value of the **measured variable.**

self-operated controller—See **controller, self-operated (regulator).**

self-regulation (inherent regulation)—The property of a **process** or machine which permits attainment of equilibrium, after a **disturbance,** without the intervention of a **controller.**

sensing element—See **element, sensing.**

sensitivity—The ratio of the change in output magnitude to the change of the input which causes it after the **steady state** has been reached.

sensor—See **transducer.**

servomechanism—An automatic **feedback control device** in which the controlled variable is mechanical position or any of its time derivatives.

setpoint—An input variable which sets the desired value of the controlled variable.

settling time—See **time, settling.**

shared-time control—See **control, shared-time.**

signal—Physical variable, one or more **parameters** of which carry information about another variable (which the signal represents).

signal, actuating error—The **reference-input signal** minus the **feedback signal.**

signal, analog—A **signal** representing a variable which may be continuously observed and continuously represented.

signal converter—See **signal transducer.**

signal, digital—Representation of information by a set of discrete values in accordance with a prescribed law. These values are represented by numbers.

signal, error—In a **closed loop,** the **signal** from its corresponding **input signal.**

signal, feedback—The return signal which results from a measurement of the directly **controlled variable.**

signal, feedforward—See **control, feedforward.**

signal, input—A **signal** applied to a **device, element,** or **system.**

signal, measured—The electrical, mechanical, pneumatic, or other variable applied to the input of a **device.** It is the analog of the **measured variable** produced by a **transducer** (when such is used.)

signal, output—A **signal** delivered by a **device, element,** or **system.**

signal, reference-input—One external to a control loop, serving as the standard of comparison for the **directly controlled variable.**

signal, return—In a closed loop, the **signal** resulting from a particular **input signal,** and transmitted by the loop and to be subtracted from the **input signal.** See also **signal, feedback.**

signal selector—A **device** which automatically selects either the highest or the lowest input **signal** from among two or more **input signals.**

signal-to-noise ratio—Ratio of **signal** amplitude to **noise** amplitude.

signal transducer (signal converter)—A **transducer** which converts one standardized transmission **signal** to another.

span—The algebraic difference between the upper- and lower-range values.

span shift—Any change in slope of the input-output curve.

static gain—See **gain, static.**

steady state—A characteristic of a condition, such as a value, rate, periodicity, or amplitude exhibiting only negligible change over an arbitrary long period of time.

steady-state deviation—See **deviaton, steady-state.**

step response—See **response, step.**

step response time—See **time, step response.**

summing point—Any point at which **signals** are added algebraically.

supervisory control—See **control, supervisory.**

system, control—See **control system.**

system, controlled—The collective functions performed in and by the equipment in which the variable(s) is (are) to be controlled.

system, controlling—(1) Of a feedback control system, that portion which compares functions of a **directly controlled variable** and a **setpoint,** and adjusts a **manipulated variable** as a function of the difference. It includes the **reference-input elements; summing point; foward** and **final controlling elements,** and **feedback elements** (including **sensing element**). (2) Of a **control system** without **feedback,** that portion which manipulates the **controlled system.**

system, directly controlled—The body, **process,** or machine directly guided or restrained by the **final controlling element** to achieve a prescribed value of the **directly controlled variable.**

system, indirectly controlled—The portion of the **controlled system** in which the indirectly controlled variable is changed in response to changes in the **directly controlled variable.**

system, linear—One of which the **time response** to several simultaneous inputs is the sum of their independent **time responses.**

time constant—The value τ in an exponential response term.

   Note: For the output of a first-order system forced by a step or an impulse, τ is the time required to complete 63.2% of the total rise or decay; at any instant during the process. τ is the quotient of the instantaneous ratio of change divided into the change still to be completed. In higher-order systems, there is a time constant for each of the first-order components of the process.

time, correction—See **time, settling.**

time, dead—The interval of time between initiation of an input change or stimulus and the start of the resulting observable response.

time, derivative action—In **proportional plus derivative control action,** for a unit ramp **signal** input, the advance in time of the **output signal** (after **transients** have subsided) caused by **derivative control action,** as compared to the **output signal** due to **proportional control action** only.

time proportioning control—See **control, time proportioning.**

time response—See **response, time.**

time, rise—The time required for the output of a system (other than first-order) to change from a small specified percentage (often 5 or 10) of the **steady-state** increment to a large specified percentage (often 90 to 95), either before or in the absence of overshoot.

time, settling—The time required, following the initiation of a specified stimulus to a system, for the output to enter and remain with a specified narrow band centered on its **steady-state** value.

time, step response—Of a **system** or an **element,** the time required for an output to change from an initial value to a large specified percentage of the final **steady-state** value either before or in the absence of overshoot, as a result of a step change to the input.

transducer—An **element** or **device** which receives information in the form of one quantity and converts it to information in the form of the same or another quantity.

transfer function—A mathematical, graphical, or tabular statement of the influence which a **system** or **element** has on a **signal** or action compared at input and at output terminals.

transient—The behavior of a variable during transition between two **steady states.**

transient overshoot—The maximum excursion beyond the final **steady-state** value of output as the result of an input change.

transmitter—A **transducer** which responds to a **measured variable** by means of a **sensing element,** and converts it to a standardized transmission **signal** which is a function only of the **measured variable.**

value, desired—The value of the **controlled variable** wanted or chosen.

value, ideal—The value of the indication, output or ultimately controlled variable of an idealized **device** or system.

variable, directly controlled—In a control loop, the variable the value of which is sensed to originate a **feedback signal.**

variable, indirectly controlled—A variable which does not originate a **feedback signal,** but which is related to, and influenced by, the **directly controlled variable.**

variable, manipulated—A quantity or condition which is varied as a function of the **actuating error signal** so as to change the value of the **directly controlled variable.**

velocity limit—A limit which the rate of change of a specified variable may not exceed.

velocity limiting control—See **control, velocity limiting.**

voltage, common mode—A voltage of the same polarity on both sides of a differential input relative to ground.

voltage, normal mode—A voltage induced across the input terminals of a **device.**

zero shift—Any parallel shift of the input-output curve.

zone, dead—(1) for a **multiposition controller,** a **zone** of input in which no value of the output exists. It is usually intentional and adjustable. (2) Dead zone is sometimes used to denote **dead band.**

Appendix C:
The Graphic Symbols
for Process Measurement
and Control

APPENDIX C

The Graphic Symbols for Process Measurement and Control

NOTE: For the serious student of process control, it is recommended that the ISA Standard on Instrumentation Symbols and Identification (ISA-S5.1/1975 approved by ANSI) be obtained. This Standard is much more complete than the very limited material presented in this appendix.

C-1. Instrument Line Symbols

All lines shall be fine in relation to process piping lines.

(1) Connection to process, or mechanical link, or instrument supply*

(2) Pneumatic signal [+], or undefined signal for process flow diagrams

(3) Electric signal

(4) Capillary tubing (filled system)

(5) Hydraulic signal

(6) Electromagnetic [§] or sonic signal (without wiring or tubing)

Notes

The following abbreviations are suggested to denote the types of power supply. These designations may also be applied for purge fluid supplies.

AS   Air Supply
ES   Electric Supply
GS   Gas Supply
HS   Hydraulic Supply
NS   Nitrogen Supply
SS   Steam Supply
WS   Water Supply

The power supply level may be added to the instrument supply line, e.g., AS 100, a 100-psig air supply; ES 24DC, a 24-volt direct current supply.

+The pneumatic signal symbol applies to a signal using any gas as the signal medium. If a gas other than air is used, the gas shall be identified by a note on the signal symbol or otherwise.

§Electromagnetic phenomena include heat, radio waves, nuclear radiation, and light.

C-2. Identification

Instruments are identified by a system of letters and numbers as shown in Table C-1. The number is generally common to all instruments of the loop of which it is a part.

The first two letters identify the function of the instrument and are selected from Table C-2. The succeeding numbers and letters identify the particular loop.

The first letter designates the measured or initiating variable such as temperature, level, flow, etc. Modifying letters such as *D* for differential, *F* for ratio, and *Q* for totalizing may follow the first letter. For example, a *TDI* is a differential temperature indicator and a *FQR* is a flow recorder with an integrator in the loop.

The succeeding letters designate one or more functions of the loop such as readout, passive function, or output.

The loop identification method assigns a number to each loop. It may begin with 1, 201, or 1201, which may include a plant area coding system.

Prefix numbers may be assigned to a number to designate plant areas. For example, 6-TRC-2 may indicate an instrument located in Plant Area 6.

If an instrument is common to more than one loop it may be assigned a separate number.

For loops that have more than one instrument with the same functional identification, suffixes should be added to the loop number; e.g., FV-2A, FV-2B, FV-2C, etc., or TE-25-1, TE-25-2, TE-25-3, etc.

Generally items such as steam traps, pressure gages, and temperature wells that are purchased in bulk quantities are not identified as loops.

Table C-1					
T	**R**	**C**	**-**	**2**	**A**
First Letter	Succeeding Letters			Loop Number	Suffix
Functional Identification			Loop Identification		
Instrument Identification or Tag Number					

TABLE C-2
MEANINGS OF IDENTIFICATION LETTERS

This table applies only to the functional identification of instruments. Numbers in table refer to notes following.

	FIRST LETTER		SUCCEEDING LETTERS (3)		
	MEASURED OR INITIATING VARIABLE (4)	MODIFIER	READOUT OR PASSIVE FUNCTION	OUTPUT FUNCTION	MODIFIER
A	Analysis (5)		Alarm		
B	Burner Flame		User's Choice (1)	User's Choice (1)	User's Choice (1)
C	Conductivity (Electrical)			Control (13)	
D	Density (Mass) or Specific Gravity	Differential (4)			
E	Voltage (EMF)		Primary Element		
F	Flow Rate	Ratio (Fraction) (4)			
G	Gauging (Dimensional)		Glass (9)		
H	Hand (Manually Initiated)				High (7, 15, 16)
I	Current (Electrical)		Indicate (10)		
J	Power	Scan (7)			
K	Time or Time-Schedule			Control Station	
L	Level		Light (Pilot) (11)		Low (7, 15, 16)
M	Moisture or Humidity				Middle or Inter-mediate (7, 15)
N (1)	User's Choice		User's Choice	User's Choice	User's Choice
O	User's Choice (1)		Orifice (Restriction)		
P	Pressure or Vacuum		Point (Test Connection)		
Q	Quantity or Event	Integrate or Totalize (4)			
R	Radioactivity		Record or Print		
S	Speed or Frequency	Safety (8)		Switch (13)	
T	Temperature			Transmit	
U	Multivariable (6)		Multifunction (12)	Multifunction (12)	Multifunction (12)
V	Viscosity			Valve, Damper, or Louver (13)	
W	Weight or Force		Well		
X (2)	Unclassified		Unclassified	Unclassified	Unclassified
Y	User's Choice (1)			Relay or Compute (13,14)	
Z	Position			Drive, Actuate or Unclassified Final Control Element	

Note: Numbers in parentheses refer to specific explanatory notes on the next two pages.

NOTES FOR TABLE C-2
MEANINGS OF IDENTIFICATION LETTERS

1. A *user's choice* letter is intended to cover unlisted meanings that will be used repetitively in a particular project. If used, the letter may have one meaning as a first-letter and another meaning as a succeeding-letter. The meanings need be defined only once in a legend, or otherwise, for that project. For example, the letter N may be defined as *modulus of elasticity* as a first-letter and *oscilloscope* as a succeeding-letter.

2. The *unclassified* letter, X, is intended to cover unlisted meanings that will be used only once or to a limited extent. If used, the letter may have any number of meanings as a first-letter and any number of meanings as a succeeding-letter. Except for its use with distinctive symbols, it is expected that the meanings will be defined outside a tagging balloon on a flow diagram. For example, *XR-2* may be a *stress recorder, XR-3* may be a *vibration recorder,* and *XX-4* may be a *stress oscilloscope.*

3. The grammatical form of the succeeding-letter meanings may be modified as required. For example, *indicate* may be applied as *indicator* or *indicating, transmit* as *transmitter* or *transmitting,* etc.

4. Any first-letter, if used in combination with modifying letters D (differential), F (ratio), or Q (integrate or totalize), or any combination of them, shall be construed to represent a new and separate measured variable, and the combination shall be treated as a first-letter entity. Thus, instruments *TDI* and *TI* measure two different variables, namely, differential-temperature and temperature. These modifying letters shall be used when applicable.

5. First-letter A for *analysis* covers all analyses that are not listed in Table C-2 and are not covered by a *user's choice* letter. It is expected that the type of analysis in each instance will be defined outside a tagging balloon on a flow diagram.†

6. Use of first-letter U for *multivariable* in lieu of a combination of first-letters is optional.

7. The use of modifying terms *high, low, middle or intermediate,* and *scan* is preferred, but optional.

8. The term *safety* shall apply only to emergency protective primary elements and emergency protective final control elements. Thus, a self-actuated valve that prevents operation of a fluid system at a higher-than-desired pressure by bleeding fluid from the system shall be a back-pressure-type *PCV,* even if the valve were not intended to be used normally. However, this valve shall be a *PSV* if it were intended to protect against emergency conditions—i.e., conditions that are hazardous to personnel or equipment, or both, and that are not expected to arise normally.

   The designation *PSV* applies to all valves intended to protect against emergency pressure conditons regardless of whether the valve construction and more of operation place them in the category of the safety valve, relief valve, or safety-relief valve.

9. Passive function *glass* applies to instruments that provide an uncalibrated direct view of the process.

10. The term *indicate* applies only to the readout of an actual measurement. It does not apply to a scale for manual adjustment of a variable if there is no measurement input to the scale.

---

† *Readily recognized self-defining symbols such as pH, O_2, and CO have been used optionally in the past in place of first-letter A. This practice may cause confusion and misunderstanding particularly when the designations are printed by machines that use only uppercase letters.*

11. A *pilot light* that is part of an instrument loop shall be designated by a first-letter followed by succeeding-letter L. For example, a *pilot light* that indicates an expired time period may be tagged KL. However, if it is desired to tag a *pilot light* that is not part of a formal instrument loop, the *pilot light* may be designated in the same way or alternatively by a single letter L. For example, a running light for an electric motor may be tagged either EL, assuming that voltage is the appropriate measured variable, or XL, assuming that the light is actuated by auxiliary electric contacts of the motor starter, or simply L.

The action of a *pilot light* may be accompanied by an audible signal.

12. Use of succeeding-letter U for *multifunction* instead of a combination of other functional letters is optional.

13. A device that connects, disconnects, or transfers one or more circuits may be either a *switch*, a *relay*, an on-off *controller*, or a *control valve*, depending on the application.

If the device manipulates a fluid process stream and is not a hand-actuated, on-off block valve, it shall be designated as a *control valve*. For all applications other than fluid process streams, the device shall be designated as follows:

A *switch*, if it is actuated by hand.

A *switch* or an on-off *controller* if it is automatic and is the first such device in a loop. The term *switch* is generally used if the device is used for alarm, pilot light, selection, interlock, or safety. The term *controller* is generally used if the device is used for normal operating control.

A *relay*, if it is automatic and is not the first such device in a loop, i.e., it is actuated by a *switch* or an on-off *controller*.

14. It is expected that the functions associated with the use of succeeding-letter Y will be defined outside a balloon on a flow diagram when it is convenient to do so. This need not be done when the function is self-evident, as for a solenoid valve in a fluid signal line.

15. Use of modifying terms *high, low,* and *middle or intermediate* shall correspond to values of the measured variable, not of the signal, unless otherwise noted. For example, a high-level alarm derived from a reverse-acting level transmitter signal shall be an LAH even though the alarm is actuated when the signal falls to a low value. The terms may be used in combinations as appropriate.

16. The terms *high* and *low*, when applied to positions of valves and other open-close devices, are defined as follows: *high* denotes that the valve is in or approaching the fully open position, and *low* denotes in or approaching the fully closed position.

Note: Words italicized correspond to entries in Table C-2.

C-3. Functional Designation of Relays

Table C-3

The function designations associated with relays may be used as follows, individually or in combination (see Table C-2, note 14). The use of a box enclosing a symbol is optional; the box is intended to avoid confusion by setting off the symbol from other markings on a diagram.

Symbol	Function
1. 1-0 or ON-OFF	Automatically connect, disconnect, or transfer one or more circuits provided that this is not the first such device in a loop (see Table C-2, note 13).
2. Σ or ADD	Add or totalize (add and subtract)†
3. Δ or DIFF.	Subtract†
4. $\pm$ + $\boxminus$	Bias*
5. AVG.	Average
6. % or 1:3 or 2:1 (typical)	Gain or attenuate (input:output)*
7. $\boxed{\times}$	Multiply†
8. $\div$	Divide†
9. $\boxed{\sqrt{}}$ or SQ. RT.	Extract square root
10. x^n or $x^{1/n}$	Raise to power
11. f(x)	Characterize
12. 1:1	Boost
13. $\boxed{>}$ OR HIGHEST (MEASURED VARIABLE)	High-select. Select highest (higher) measured variable (not signal, unless so noted).
14. $\boxed{<}$ OR LOWEST (MEASURED VARIABLE)	Low-select. Select lowest (lower) measured variable (not signal, unless so noted).
15. REV.	Reverse
16.	Convert
a. E/P or P/I (typical)	For input/output sequences of the following: Designation Signal E Voltage H Hydraulic I Current (electrical) O Electrmagnetic or sonic P Pneumatic R Resistance (electrical)
b. A/D or D/A	For input/output sequences of the following: A Analog D Digital
17. $\int$	Integrate (time integral)
18. D or d/dt	Derivative or rate
19. 1/D	Inverse derivative
20. As required	Unclassified

Used for single-input relay.
†*Used for relay with two or more inputs.*

C-4. Special Abbreviations

Table C-4

FOR ABBREVIATIONS OTHER THAN INSTRUMENT IDENTIFICATION LETTERS OF TABLE C-2

ABBREVIATION	MEANING
A	Analog signal
ADAPT.	Adaptive control mode
AS	Air supply
AVG.	Average
C	Patchboard or matrix board connection
D	⎰ Derivative control mode ⎱ Digital signal
DIFF.	Subtract
DIR.	Direct-acting
E	Voltage signal
ES	Electric supply
FC	Fail closed
FI	Fail indeterminate
FL	Fail locked
FO	Fail open
GS	Gas supply
H	Hydraulic signal
HS	Hydraulic supply
I	⎰ Current (electrical) signal ⎱ Interlock
M	Motor actuator
MAX.	Maximizing control mode
MIN.	Minimizing control mode
NS	Nitrogen supply
O	Electromagnetic or sonic signal
OPT.	Optimizing control mode
P	⎰ Pneumatic signal ⎨ Proportional control mode ⎱ Purge or flushing device
R	⎰ Automatic-reset control mode ⎨ Reset of fail-locked device ⎱ Resistance (signal)
REV.	Reverse acting
RTD	Resistance (-type) temperature detector
S	Solenoid actuator
S.P.	Setpoint
SQ.RT.	Square root
SS	Steam supply
T	Trap
WS	Water supply
X	⎰ Multiply ⎱ Unclassified actuator

C-5. General Instrument Symbols

Balloons

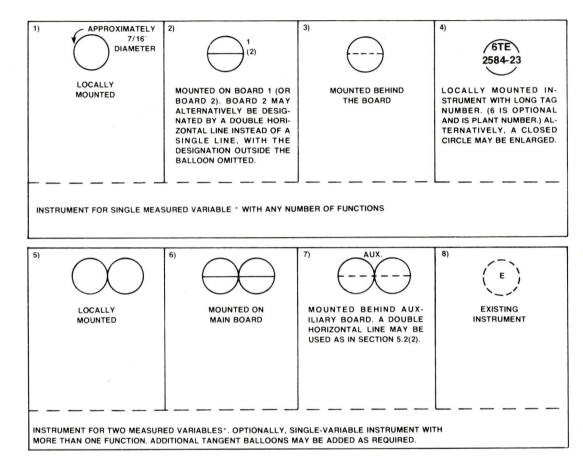

CONTROL VALVE BODIES

| 1) GLOBE, GATE, OR OTHER IN-LINE TYPE NOT OTHER-WISE IDENTIFIED | 2) ANGLE | 3) BUTTERFLY, DAMPER, OR LOUVER | 4) ROTARY PLUG OR BALL |
| 5) THREE-WAY | 6) ALTERNATIVE 1 | 7) ALTERNATIVE 2 | |

FOUR-WAY

ACTUATORS

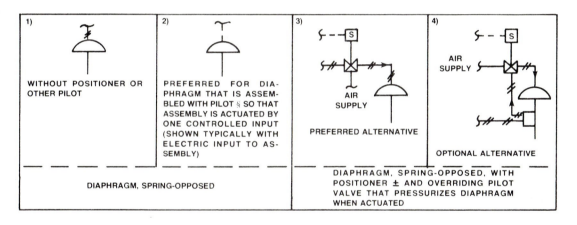

| 1) WITHOUT POSITIONER OR OTHER PILOT | 2) PREFERRED FOR DIA-PHRAGM THAT IS ASSEM-BLED WITH PILOT § SO THAT ASSEMBLY IS ACTUATED BY ONE CONTROLLED INPUT (SHOWN TYPICALLY WITH ELECTRIC INPUT TO AS-SEMBLY) | 3) AIR SUPPLY / PREFERRED ALTERNATIVE | 4) AIR SUPPLY / OPTIONAL ALTERNATIVE |

DIAPHRAGM, SPRING-OPPOSED

DIAPHRAGM, SPRING-OPPOSED, WITH POSITIONER ± AND OVERRIDING PILOT VALVE THAT PRESSURIZES DIAPHRAGM WHEN ACTUATED

SELF-ACTUATED REGULATORS, VALVES, AND OTHER SERVICES

FLOW

1) AUTOMATIC REGULATOR WITH INTEGRAL FLOW INDICATION. TAG REGULATOR FCV-5 IF IT DOES NOT HAVE INTEGRAL FLOW INDICATION.

2) (UPSTREAM ALTERNATIVE) (DOWNSTREAM ALTERNATIVE) INDICATING ROTAMETER WITH INTEGRAL MANUAL THROTTLE VALVE

HAND

1) HAND CONTROL VALVE IN PROCESS LINE

2) HAND-ACTUATED ON-OFF SWITCHING VALVE IN PNEUMATIC SIGNAL LINE

3) MANUALLY ADJUSTABLE RESTRICTION ORIFICE IN SIGNAL LINE

LEVEL

1) LEVEL REGULATOR WITH MECHANICAL LINKAGE

PRESSURE

1) PRESSURE-REDUCING REGULATOR, SELF-CONTAINED

2) PRESSURE-REDUCING REGULATOR WITH EXTERNAL PRESSURE TAP

3) DIFFERENTIAL-PRESSURE-REDUCING REGULATOR WITH INTERNAL AND EXTERNAL PRESSURE TAPS

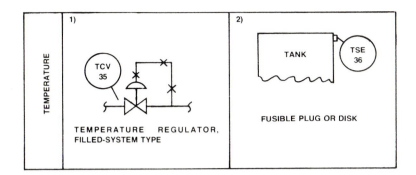

Primary Elements

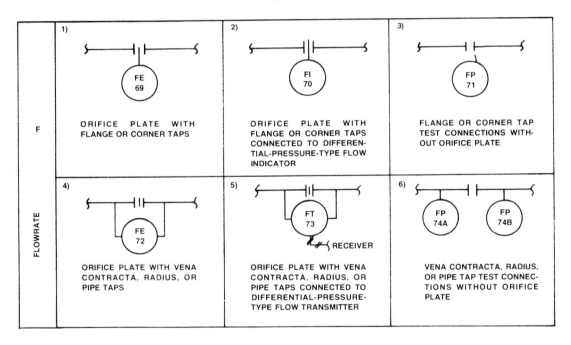

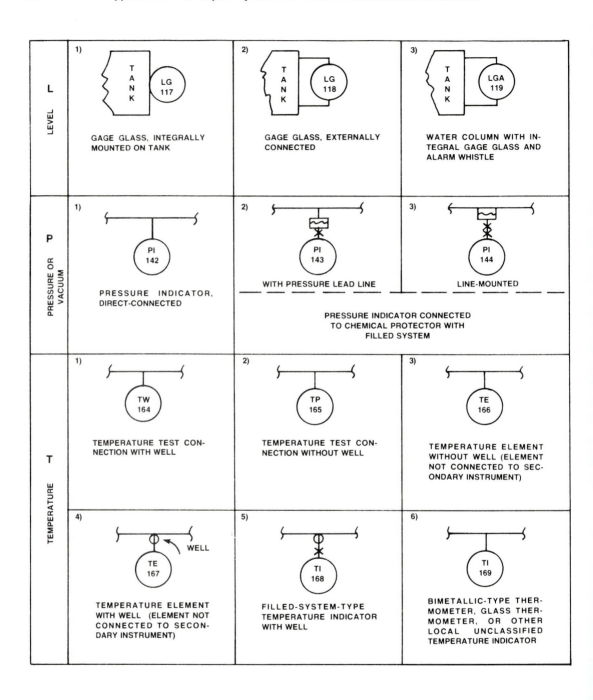

Appendix D:
Solutions to All Exercises

APPENDIX D

Solutions to All Exercises

Chapter 2

Exercise 2-1.

Controlled variable—oven temperature
Manipulated variable—electric current
Disturbances—ambient temperature
        oven contents
        endothermic/exothermic reactions
        leaks
        door opening/closing, etc.

Exercise 2-2.

Controlled variable—water temperature
Manipulated variable—gas flow
Disturbances—water usage
        Btu content of gas
        ambient temperature
        inlet water temperature, etc.

Exercise 2-3.

Manual System:
1. Take a sample of pool water
2. Use litmus paper or equivalent to measure pH
3. Add acidic solution as necessary
   In the above system: controlled variable—pool water pH
   Manipulated variable—acidic solution added
   Disturbances—variation in strength of acidic solution
               evaporation rate
               pH of any make-up water added to pool, etc.

Exercise 2-4.

A feedback automatic control system might appear as follows:

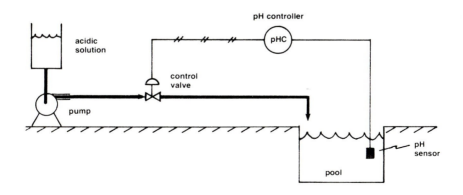

Exercise 2-5.

Controlled variable—intake manifold vacuum
Manipulated variable—accelerator (adjusted by a transducer-servomotor that drives a chain)
Disturbances—grade of road
                weight
                engine performance, etc.

Exercise 2-6.

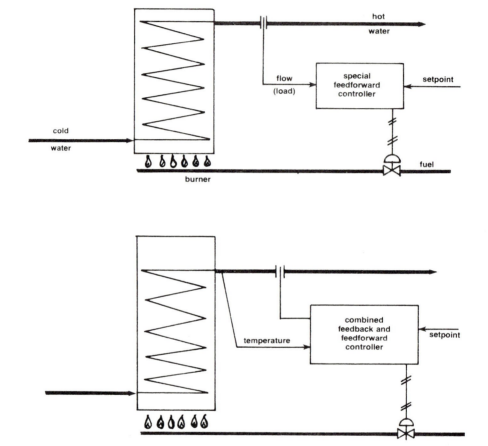

Exercise 2-8.

A "moonshot" is based on a predetermination of flight path (trajectory). In order to arrive at a specific point in space on the surface of the moon, the calculations involve consideration of the gravitational effects on all bodies, motions of all bodies, rocket thrust, rocket attitude, etc. An elaborate computer model is used to determine flight trajectory.

A mid-course correction is needed to correct for inadequacies in the model and in the less-than-perfect performance of all the hardware components of the rocket system. The mid-course correction is a feedback adjustment to eliminate errors between desired and actual flight trajectory.

Exercise 2-9.

Consider, for example, a small home thermostat that allows input of a normal maximum desired temperature, e.g., 78°F, and the input of a normal minimum desired temperature, e.g., 68°F. Various wider limits can be in effect during the evening hours while the occupants are asleep, e.g., 80°F and 65°F. Also, during the day when the occupants are away at work or away on a trip, wider bands may also be programmed. A built-in clock can be included in the thermostat case so that room temperature can be brought back within normal control limits at prescribed times. This type of management of home temperature can be programmed on a routine seven-day week clock, and various subroutines adjusted as necessary to reflect the travel and living patterns of the occupants. Of course, one portion of the thermostat's operation is to activate the necessary heating or air conditioning systems when the actual temperature varies outside of the desired band width.

Chapter 3

Exercise 3-1.

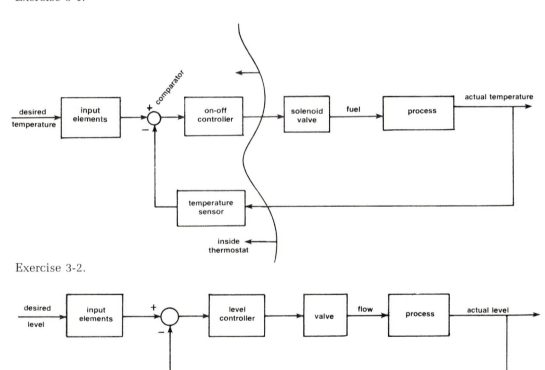

Exercise 3-2.

Exercise 3-3.

$$\frac{12 \text{ psi}}{200°F} = .06 \text{psi per}°F$$

Exercise 3-4.

$$\frac{16\text{ma}}{60\text{GPM}} = 0.266\text{ma per GPM}$$

Exercise 3-5.

Input elements—no
Electronic transmission—no
Large pneumatic valve—probably yes
2,000 ft. pneumatic transmission system—yes
Pneumatic controller—no
Electronic controller—no
Bare thermocouple—no
Orifice meter—no
Chromatograph—yes, probably
Thermocouple in thermowell—yes, probably

Exercise 3-6.

Liquid flow—very rapid
Gaseous flow—less rapid than liquid flow
Liquid level in a small tank—moderately rapid
Liquid pressure in an enclosed tank—rapid
Gaseous pressure in a large tank—slow
Composition in a large distillation column—slow
Temperature in a liquid filled tank—slow
Your co-workers—you be the judge!

Chapter 4

Exercise 4-1.

The distinctions between transmission to a board-mounted (central control room) controller and to a board indicator and/or recorder are minor; the transmission systems are virtually identical. The only difference is in the usage of the signal once it gets to the control room.

Exercise 4-2.

The thermowell causes more lag and slows down the temperature response. The curve in Fig. 4-2 might be more S-shaped if a thermowell is used:

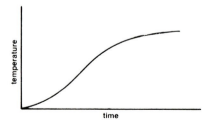

Exercise 4-3.

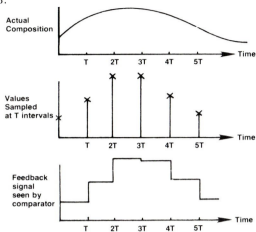

Exercise 4-4.

If there are errors in accuracy, this will not introduce any instability or disturbance into the process, but may cause the operator to need to readjust his setpoint; this is a minor irritant but not nearly so serious as instabilities caused by lack of reproducibility.

Exercise 4-5.

Refer to Fig. 4-8.
A time constant of 3 seconds dictates a maximum pneumatic transmission distance of 700'.

Exercise 4-6.

Both transmission systems are of equivalent importance insofar as their dynamics are concerned.

Chapter 5

Exercise 5-1.

a. 0.5
b. 500%

Exercise 5-2.

At steady state, there is an error. This error produces a controller output which produces a certain amount of manipulated variable. All of this manipulated variable is used to maintain the current state, and there is no net or incremental amount of manipulated variable available to produce a change from the current steady state. In effect, the manipulated variable output is consumed in maintaining the status quo.

Exercise 5-3.

It makes no difference whatsoever if the error is caused by a change in disturbance. The dynamics of the response will be similar either way; of course, the transient response of the process will lead to a new final value if there has been a change in setpoint.

Exercise 5-4.

If manual reset is used to bias the controller output to eliminate offset at a specific operating point, there will be no offset in evidence until there is either a change to a new setpoint or there is a change in some of the disturbances at work on the system. Since disturbances change frequently, and setpoints often change, the elimination of offset by manual reset is not a long-term operating solution.

Exercise 5-5.

The proposed controller would have good dynamic response for large errors and when the error is small, the reset-only controller could be used to trim out offset. The significant advantage would be that the dynamic lag penalties inherent with reset would not be present except when that mode is in active use.

Exercise 5-6.

The two algorithms are equivalent to one another but the noninteracting form is a bit easier to tune. Because all of the modes are independent of one another, when you change the gain of one mode, it does not produce a change in the gain of the other modes.

Exercise 5-7.

Filtering is necessary on the rate mode because otherwise very small—but rapid—changes in the error signal might produce very large outputs from the rate mode. Such would be deleterious to loop dynamics.

Exercise 5-8.

Proportional

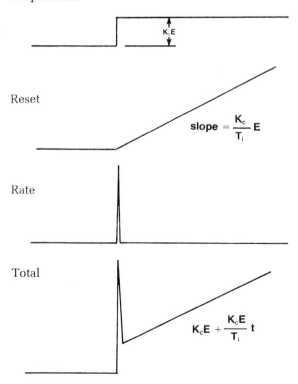

Reset

$$\text{slope} = \frac{K_c}{T_i} E$$

Rate

Total

$$K_c E + \frac{K_c E}{T_i} t$$

Exercise 5-9.

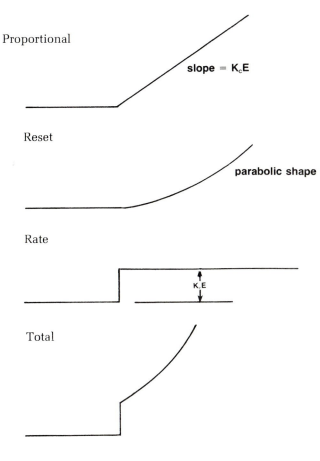

Proportional

slope = $K_c E$

Reset

parabolic shape

Rate

$K_c E$

Total

Chapter 6

Exercise 6-1.

A 0.5 psi air-to-valve signal increase produces a flow increase of 15%; therefore, at a base flow of 20 GPM, the increase is 3 GPM.

Exercise 6-2.

For a linear valve the gain of the valve does not change; therefore, at 20 GPM, a 0.5 psi increase produces a change of 1.5 GPM.

Exercise 6-3.

Flow, $M_L = C_v \sqrt{\dfrac{\Delta P}{G}}$

$= 20 \sqrt{\dfrac{10}{1}}$

$= (20)\,(3.17)$

$= 63.4 \text{ GPM}$

Exercise 6-4.

Flow, $M_g = 63.3\ C_v\gamma\ \sqrt{\Delta PV_1}$
$= (63.3)(32)(0.8)\ \sqrt{(10)(3.103)}$
$= (1620)\ \sqrt{31.03}$
$= (1620)(5.57)$
$= 9{,}023$ lbs per hour

Exercise 6-5.

Refer to Figure 6-8. With 10% of the dynamic drop across the value and in the range of 50 to 70% opening, the valve characteristic curve is effectively linear.

Exercise 6-6.

For most situations we prefer for a control valve to fail closed, unless it is necessary to maintain flow for safety considerations.

Chapter 7

Exercise 7-1.

Curve 1—Time constant of approximately 1 second
Curve 2—Time constant of approximately 3.5 seconds
Curve 3—Time constant of approximately 7.5 seconds

Exercise 7-2.

A time constant of 0 implies an algebraic response, i.e., instantaneous correspondence of output to input.

A time constant of infinity implies no response whatsoever.

Exercise 7-3.

Time constant $=$ capacitance/conductance
$= 28\text{ft }^3/6\text{ ft}^3$ per min $= 4.6$ min

Exercise 7-4.

Time constant $=$ capacitance/conductance
$= \dfrac{(0.26\text{ lbs })(0.6\text{ Btu/lb }/^\circ\text{F})}{(3.7\text{ Btu/ft }^2/^\circ\text{F})(0.16\text{ ft }^2)}$
$= 0.264$ min

Exercise 7-5.

Dead time $= 3.6$

Exercise 7-6.

Dead time $=$ distance/velolcity
$= \dfrac{16\text{ft}}{21\text{ ft /min}} = 0.762$ min

Exercise 7-7.

As the controller gain is increased, the response of the loop becomes more dynamic and much more rapid; there is an increase in the speed of response. This increased speed tends to make the loop less stable. Finding the optimum controller gain implies making a trade-off between increased speed of response and decreased stability.

Exercise 7-8.

As more and more dead time is introduced, the loop becomes less and less stable. During the dead time of a component, there is no output and, thus, no corrective action is initiated.

Chapter 8

Exercise 8-1.

For two-mode and three-mode controllers, there can be many response curves with a decay-ratio of one-quarter but with oscillations at different frequencies.

Exercise 8-2.

Tuning parameters are not a function of whether a loop is a liquid level loop, a flow loop, etc., but tuning constants depend upon the individual dynamic parameters and the individual gains (or sensitivities) of the various components that make up the loop. Since these vary from one loop to the next, the desirable tuning parameters vary from loop to loop.

Exercise 8-3.

Proportional only
$$K_c = 0.5 \, S_u = (0.5)(0.3 \text{ psi/ft }) = 0.15 \text{ psi/ft}$$

Proportional-plus-reset
$$K_c = 0.45 \, Su = (0.45)(0.3 \text{ psi/ft }) = 0.135 \text{ psi/ft}$$
$$T_i = P_u/1.2 = 3/1.2 = 2.5 \text{ min}$$

Proportional-plus-derivative
$$K_c = 0.6 \, S_u = (0.6)(0.3) = 0.18 \text{ psi/ft}$$
$$T_d = P_u/8 = 3/8 = 0.375 \text{ min}$$

Proportional-plus-reset-plus-derivative
$$K_c = 0.6 \, S_u = 0.18 \text{ psi/ft}$$
$$T_i = 0.5 \, P_u = 1.5 \text{ min}$$
$$T_d = P_u/8 = 0.375 \text{ min}$$

Exercise 8-4.

The general units of K are the units of the controller input divided by controller output. The product of all of the gains around the feedback loop KK_c is unitless.

Exercise 8-5.

Proportional
$$K_c = 1/L_r R_r = 1/(0.15 \text{ min })(0.6 \text{ ft /psi-min })$$
$$= 11.11 \text{ psi/ft}$$

Proportional-plus-reset
$$K_c = 0.9/L_r R_r = 0.9/(0.15)(0.6)$$
$$= 10 \text{ psi/ft}$$
$$T_i = 3.33 \, L_r = (3.33)(0.15 \text{ min })$$
$$= 0.499 \text{ min}$$

Proportional plus-reset-plus-rate
$K_c = 1.2/L_rR_r = 13.3$ psi/ft
$T_i = 2.0\ L_r = 0.3$ min
$T_d = 0.5L_r = 0.075$ min

Exercise 8-6.

Proportional
$PB = (1/K_c)(100)$
$= (0.09)(100) = 9\%$

Proportional-plus-reset
$PB = (1/K_c)(100)$
$= 10\%$
$RPM = 2.004$ repeats per min

Proportional-plus-reset-plus-rate
$PB = 7.5\%$
$RPM = 3.33$
$T_d = 0.075$ min

Chapter 9

Exercise 9-1.

Superficial inspection of Figure 9-2 indicates that a change in T_i will, of course, change the temperature of the reacting mass and this will be sensed by the temperature bulb. Since corrective action must pass through an additional controller, it is tempting to speculate that corrective action will take longer. This is not true. By the creation of the interior loop around part of the lag of the process, that portion of the process is "speeded up" and the net effects of corrective action are faster. In effect, the implementation of cascade control, properly tuned, will improve response to disturbances no matter whether they enter the inner loop or the outer loop.

Exercise 9-2.

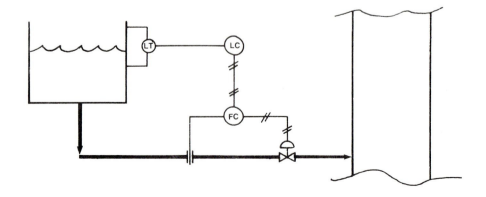

With the cascade arrangement, any change in flow through the valve—whether caused by a change in level in the tank or operating pressure within the column or for whatever reason—will be immediately sensed by the orifice meter and corrective action taken directly. The net result is much steadier liquid flow rate into the column.

Exercise 9-3.

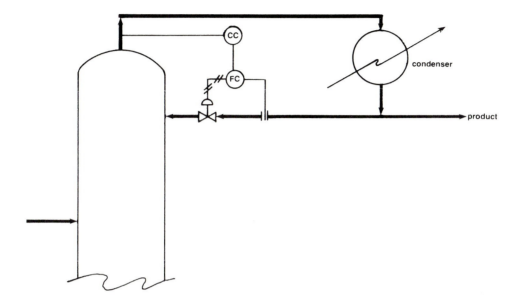

Exercise 9-4.

Controlled gin stream:

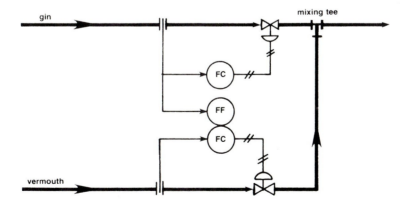

Uncontrolled gin stream:

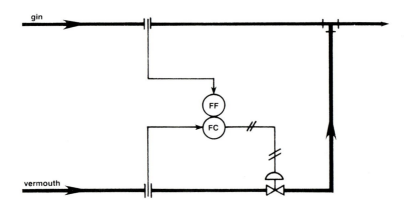

Exercise 9-5.

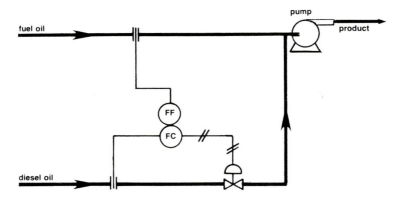

Exercise 9-6.

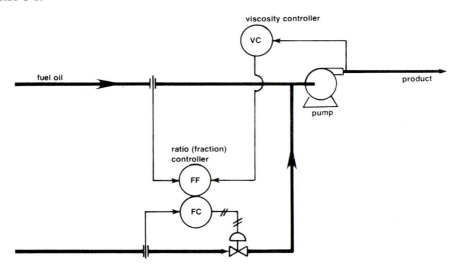

Exercise 9-7.

The sensor (viscosimeter) should be as close to the point of intimate mixing as possible. As you move it further and further downstream you introduce a distance/velocity dead time into the feedback path and the further downstream the sensor is moved, the larger the dead time becomes. As this dead time becomes more and more significant, it becomes more and more difficult to operate the feedback control system.

Chapter 10

Exercise 10-1.

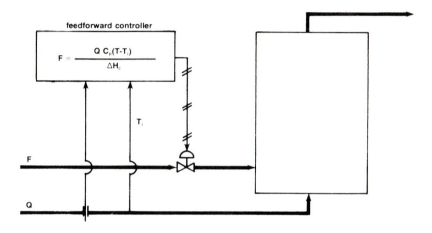

Exercise 10-2.

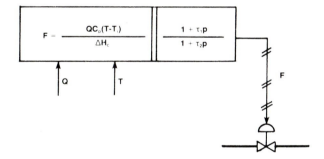

Tuning adjustment can vary γ_1/γ_2 to adjust the dynamic compensation of F.

Exercise 10-3.

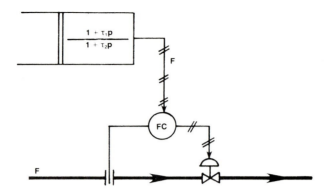

Exercise 10-4.

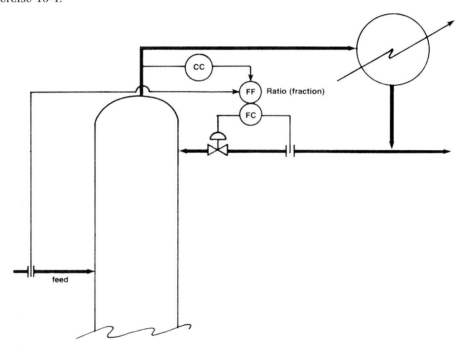

Exercise 10-5.

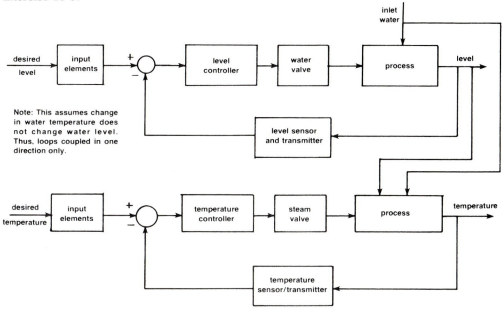

Note: This assumes change in water temperature does not change water level. Thus, loops coupled in one direction only.

Exercise 10-6.

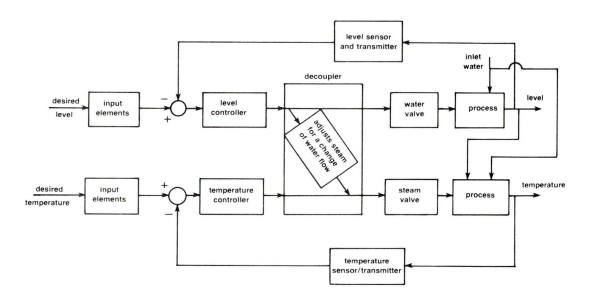

Chapter 11

Exercise 11-1.

Digital computer technology allows us to process information and to easily communicate the results. Consequently, control and management can be automated to an extent not heretofore practical.

Exercise 11-2.

In many cases we need to log data and to view such data as a portion of the output of the process. In a pilot plant operation, for example, the product chemicals or product streams produced may be insignificant as compared to the information and data produced through plant operation.

Exercise 11-3.

Different types of loops have different amounts of process lag. Generally, flow loops have the least and temperature loops have the most, with level and pressure loops being intermediate. The faster a loop, the more rapid it must be sampled in DDC.

Exercise 11-4.

In effect, if you sample and hold a signal as shown, you effectively introduce a delay that is equivalent to a dead time of one-half the sample time. This is shown:

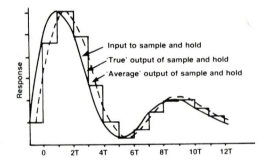

Exercise 11-5.

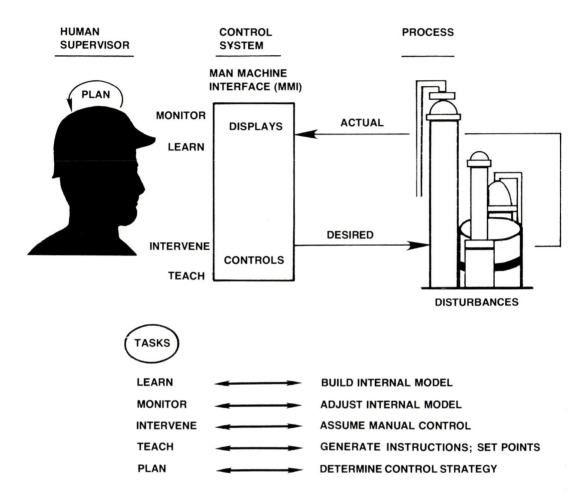

THE TASK OF THE HUMAN SUPERVISOR

Exercise 11-6.

Optimization implies mathematical computation and the degree of such computation may vary from the most elementary to very extensive, sophisticated situations. To make these computations implies the need for computer hardware and this, in turn, requires programming capabilities for those who design and implement the process control and management programs on the computer.

Chapter 12.

Exercise 12-1.

Results will vary depending upon the unit reviewed.

Exercise 12-2.

Results will vary depending upon the unit reviewed.

Exercise 12-3.

Results will vary depending upon the particular vendors selected.

Exercise 12-4.

Some units can be handled entirely satisfactorily by analog process controllers, and in such cases, there is no need or desirability for more sophisticated digital process controllers. In some cases, however, more sophisticated computations are needed or desirable at the individual loop level, and in such cases, digital controllers are more satisfactory.

Exercise 12-5.

As operating costs and raw material costs escalate, then the necessity of and the rate of return for quality process control and process management are increased. Therefore, it becomes more and more attractive (mandatory?) to do a better control job.

Exercise 12-6.

Increasing legal and regulatory constraints and concerns with product liability and product specification significantly escalate the requirement for effective process control and process management. This requires higher and higher levels of process automation.

Index

INDEX